Eat,
Poop,
Die

Also by Joe Roman

Whale

Listed: Dispatches from America's Endangered Species Act

Eat, Poop, Die

HOW ANIMALS MAKE OUR WORLD

Joe Roman

LITTLE, BROWN SPARK
New York Boston London

Little, Brown Spark
Hachette Book Group
1290 Avenue of the Americas, New York, NY 10104
littlebrownspark.com

First Edition: November 2023

Little, Brown Spark is an imprint of Little, Brown and Company, a division of Hachette Book Group, Inc. The Little, Brown Spark name and logo are trademarks of Hachette Book Group, Inc.

The publisher is not responsible for websites (or their content) that are not owned by the publisher.

The Hachette Speakers Bureau provides a wide range of authors for speaking events. To find out more, go to hachettespeakersbureau.com or call (866) 376-6591.

Little, Brown and Company books may be purchased in bulk for business, educational, or promotional use. For information, please contact your local bookseller or the Hachette Book Group Special Markets Department at special.markets @hbgusa.com.

Illustrations by Alex Boersma

ISBN 9780316372923
LCCN 2023938980

Printing 1, 2023

LSC-C

Printed in the United States of America

To my dad, José Roman; uncle Joe Sweeney; and friend and colleague Jim McCarthy—whose atoms are now part of new constellations

Contents

Eat,
Poop,
Die

I

Beginnings

Just before dawn on November 14, 1963, the *Ísleifur 2* had set a bottom longline off the southeast coast of Iceland. Most of the crew were belowdecks—resting up before they retrieved the line and unhooked the cod—but the engineer noticed a strong smell of sulfur as he finished his morning coffee on deck. He checked the wake of the vessel. There was no sign of sewage—no cause for alarm—so he joined the other men below.

Half an hour later, the cook on watch noticed the ship starting to rock as if caught in a whirlpool. Dark smoke rose above the turquoise surface of the sea. He yelled down to the skipper. All hands now awake, they looked to see if there was a ship in distress nearby. But they saw only a plume.

Four hundred feet below, the seafloor trembled. Then the tephra—ash, cinders, and lapilli (rock fragments about the size of rabbit pellets)—spewed up from the ocean, dwarfing the fishing boat. Smoke from the explosion rose five hundred feet above the sea surface, once blue, now greenish brown. As the tephra column reached a height of more than two miles, it became obvious: the crew had set their gear near a volcanic fissure.

There were no fish on the line when they finally pulled it in from the boiling sea.

BY THE NEXT morning, a new island had risen thirty-three feet above the surface of the North Atlantic. The island continued to rise about two hundred feet per day in an uprush of magma, cinder, and ash, and within a week, the eruption column, white by day, pink at night, reached six miles into the air. Flashes of lightning creased the sky.

Inhabitants of Heimaey, the only town on the Vestmannaeyjar, Iceland's Westman Islands, reported seeing glowing embers on the horizon as seawater entered the new crater. Six large earthquakes rattled the town. On December 6, three French journalists took a speedboat from Heimaey to the new island and stayed for about fifteen minutes before an eruption chased them off.

The media attention in Iceland and abroad got people wondering what they should call this new landform. For a moment it seemed that the first person who had set eyes on the new island, the cook Ólafur Westmann, might be honored by having the island named after him: Olafsey (Olaf's Island). Others in Heimaey preferred Vesturey (West Island).

Icelanders take their names very seriously—the government still has final say over acceptable baby names in the country; there are no Lucifers, no Ariels—so the Icelandic government convened Örnefnanefnd, the place-name committee, to decide. The choice was announced on the radio, and soon after, one of Ólafur's shipmates found the cook cleaning up in the galley, dishrag in hand, on the verge of tears. "They gave it a terrible name," he muttered. "Surtsey."

The committee had turned to Norse mythology: During Ragnarök, the prophesied end of the world, the giant Surtur will bring fire to fight the god Freyr. The lava vent was a lethal red with water

4

boiling all around it, so the committee called the new landform Surtur's Island—Surtsey.

Westman Islanders, angry at not having been consulted, sailed to Surtsey's shore and erected a sign with the name VESTUREY. Surtur responded by pelting the islanders with pumice and mud. No lives were lost. Surtsey stuck.

IN ITS FIRST year, Surtsey expanded at thirty cubic yards a second, adding an area almost as big as the Great Pyramid of Giza each day. The lava plain was a glistening black with ropes of hot lava unraveling toward the sea.

Sigurdur Thórarinsson, a professor at the University of Iceland, was the first volcanologist to land on Surtsey, about three months after the initial eruption. He and a few fellow scientists were collecting geological samples along the shore when they noticed waterspouts in the ocean. Lava bombs crashed into the water and started falling around them. Each one up to a yard in diameter, the bombs landed on the beach with resounding thuds as the wet volcanic sand boiled beneath the red-hot lava. "Under such circumstances there is really only one thing to do," Thórarinsson recalled. "Suppress the urge to take to your heels and endeavor to stand still and stare up in the air, trying not to dodge the bombs until the very moment they seem to be about to land on your head." Stop and look up—but not for too long, or the soles of your boots will start to smolder. Thórarinsson noticed that the research vessel was moving farther offshore, away from the danger.

The volcanologists were soon enveloped in "warm and cozy" clouds of pumice, grains of rock so light, they floated in the air. It was hard to breathe, and visibility was down to zero, but at least the larger bombs had stopped falling. As the wind carried the pumice cloud away, Thórarinsson and his colleagues waded back to their dinghies and rowed to the ship.

No one returned to Surtsey until the vent stopped exploding.

***

WHEN THE LAVA bombs subsided, Surtsey gave biologists the rare opportunity to study life from the first days of an island's emergence. It was an "ecologist's dreamworld," according to Charlie Crisafulli, who has studied Mount Saint Helens in Washington since it erupted in 1980. Unlike that eruption—which covered forests and grasslands, so there was some residual life beneath the ash—Surtsey rose in the middle of the ocean. It was inaccessible, at first, with no animals or plants and a hostile environment. As soon as he stepped off the helicopter, he realized that Surtsey was a perfect place to study how ecological communities assembled.

"The materials coming out of volcanic events can be toxic with sulfur, chlorine, and fluoride compounds," Crisafulli told me over the phone a few years after he visited Surtsey. "This is a huge problem for animals and plants." There were too many bads (toxins) and not enough goods (nutrients) for anything to survive on Surtsey. The gases, lava, and tephra that volcanoes belch out lack many of the basic building blocks of ecosystems, such as carbon and nitrogen, but the rocks are phosphorus-rich. "What's happened in an old landscape like the one you're sitting in right now, the Green Mountains, the White Mountains, the Adirondacks"—I was talking to him from my home in Vermont—"is that phosphorus has long weathered out of those rocks," Crisafulli said. "But volcanic landscapes provide a new, fresh batch of phosphorus that can often be rendered quite readily." So there was plenty of phosphorus on Surtsey, but nitrogen, at least in the form that animals and plants could use, was quite rare. Both elements are essential to life, forming the basic building blocks of DNA and protein and helping to power mitochondria, the workhorses of the cell.

During Surtsey's first decade, there was little vegetation in the volcanic sand and lava. When it rained, the water leached through the porous lava and eventually reached the ocean; when it didn't, Surtsey was like a desert or the highlands of Iceland, which are so barren NASA once used them to train astronauts for the lunar landing. Any plants that did show up on Surtsey were faced with a shortage of nitrogen in the soil.

Nobody knew it yet, but the answer to the nutrient problem could be seen even as the vent was still erupting. A couple of kittiwakes, yellow-beaked seabirds common on the mainland, alighted on the craggy, vinyl-black shores. These birds, and the gulls and fulmars that followed, would deliver the first concentrated nitrogen in the form of uric acid, one pasty poop at a time.

THE FIRST LIFE on this new landscape arrived by sea or fell out of the sky. Tiny seeds of wind-dispersed plants—willows, orchids, and ferns—rained softly over the island. To stay airborne, these seeds travel light, carrying little in the way of food or nutrients with them. They showed up on Surtsey's inhospitable shores, but with limited resources, they either never sprouted or soon withered and disappeared.

Large, buoyant seeds were swept ashore on the ocean currents. "If you are traveling by sea," Borgthór Magnússon,* one of the island's longtime naturalists, told me, "then you can afford to pack nutrients for your establishment." The first documented species on the new island was the sea rocket, a succulent that washed up along Surtsey's sandy edge and took hold. Its seed has a cork-like covering that helps it float and keeps it safe from salt water. But lava was still flowing into the sea in a drama of magma, gray ocean, breakers, and steam. Tephra, ash, and cinders from a nearby vent buried the young plants.

In the beginning, pioneers like the sea rocket were no match for the active volcano, but they kept arriving, and the island slowly cooled. Soon, ocean-borne seeds of sea sandwort and oysterleaf washed up on the bleak edges of Surtsey's coast. The seeds of both

---

* Borgþór Magnússon is the Icelandic spelling of the name; the þ is pronounced like "th" in English, so þor is "thor." For ease of reading, I'll use the English spellings of Icelandic names.

species are built for oceanic travel, packing enough nutrients with them to take root when washed ashore. Suited to the wind and cold, the sandwort hugged the barren sands where few other species could survive. It formed a bell of bright succulent leaves aboveground; deep roots stretched out below, absorbing water and nutrients from the crevices and sand. In cross section, the sandwort resembles a Portuguese man-of-war: a green sail above the surface and long tentacles below. Decades later, sandwort still covers parts of Surtsey in striking herdlike patterns, the most fetching plants on the island.

Oysterleaf is as seaworthy as its species epithet, *maritima*, implies. You couldn't engineer a better pioneer for a new island in the North Atlantic. Oysterleaf seeds usually stay dormant until they are shocked by the cold sea—temperatures in the thirties enhance germination. By the time the seeds reached Surtsey, they were ready to grow. The oysterleaf seedlings stayed close to the boulders at the rocky edges of the island, keeping their heads out of the wind. During Surtsey's first decade or so, the plants' low-hanging flowers provided rare splashes of blue against a monochrome landscape.

Life remained sparse at first. Only the most hardy and well-provisioned plants could survive. There were no pedestrian invertebrates—no daddy longlegs, worker ants, or crickets, the usual suspects that show up on the pumice of volcanoes. A few insects blew in. The first one recorded was a migratory moth, then a couple of midges. Many of these bugs—called *fallout fauna*—probably died of fatigue, desiccation, or low temperatures. One zoologist described the first insects to arrive in a volcanic landscape as "derelicts of dispersal."

Nonetheless, slowly but surely, animals—insects, birds, then seals—made their way to the young island.

It wasn't that long ago that animals were dismissed by many scientists as bit players on the planet; plants and microbes took center stage. But in the past decade or so, there has been a radical shift in our understanding of how the world is shaped by predators and herbivores. Landmark studies of seabirds, whales, sea otters, salmon, wildebeests, bison, spiders, grasshoppers, cicadas, and other animals have shown that they can alter the landscapes and seascapes where they live, with major impacts on ecological function and the services these animals provide. Much of this remains unseen; few people realize that when they recline on the white sands of Hawaii and other tropical beaches, they are lying in the waste of parrotfish, the poop from coral meals.

Animals matter. Creatures—sometimes with fur or scales, sometimes red in tooth and claw, perhaps with talons and wings, wild and free-roaming—are a fundamental mechanism of sustaining life and a source for the nutrients it requires. It's only after thousands of years of serial depletion by our species that scientists are beginning to understand the interconnectedness of these energy transfers.

Follow the nutrients. The essential elements of carbon, nitrogen, and phosphorus move in geologic time, carried by gravity, wind, and currents. Downhill. Downwind. Downstream.

If they reach the deep sea, phosphate and ammonia molecules—the common sources of phosphorus and nitrogen—can be locked away in the ocean depths for hundreds of years unless they hit an area of upwelling, where waters are drawn to the surface. Such areas are rare in the ocean. There is another way for these vital nutrients to move thousands of feet up the water column: they can catch a ride in the belly of a whale.

Foraging sperm whales feed on giant squid and other deep-sea creatures, but they must return to the surface at least once an hour for a postprandial breath of air; there, they rest, digest, and often release enormous fecal plumes rich in phosphates, nitrogen, and iron. The nutrients in the plumes can be picked up by phytoplankton (also known as microalgae) and consumed by zooplankton, such as krill or tiny copepods. The krill or their fish predators might then be consumed by seabirds—gulls, fulmars, terns, penguins, petrels, shearwaters, albatrosses, boobies, and magnificent frigate birds—and airlifted to their breeding grounds. Back at their nests, the birds feed their young by vomiting up their sea meals and excrete nitrogen-rich uric acid—the striking white paste that is released along with the feces—onto the land.

We can follow these elements from the deep sea to the coasts, rivers, forests, savannas, and mountains of the world. A geological journey that would take thousands or millions of years—the tectonic plates beneath Iceland move at a rate of approximately an inch and a half per year, about the speed of growing fingernails—can be reversed in a single dive, a short flight back to a barren rock, and a Pollock-like spatter.

Animals are the beating heart of the planet. In the same way that trees work as the Earth's lungs—inhaling carbon dioxide and exhaling oxygen—animals pump nitrogen and phosphorus from deep-sea gorges up to mountain peaks and across hemispheres from the poles to the tropics. Trillions of animals live the traveling life—they fly, run, swim, walk, even dig. Large and medium-size animals—whales, elephants, bison, salmon, and seabirds—can move nutrients hundreds and sometimes thousands of miles, across oceans, streams, mountains, valleys, prairies, and remote volcanic islands. These long-distance travelers are the world's arteries. Cicadas, midges, krill, and other invertebrates, if we take this idea a step further, are the capillaries, delivering nutrients to Earth's tissues.

It's not just poop and carcasses. Animals change the world through their consumption too. They eat plants. They eat plant-eaters. They change the chemistry of the world just by instilling fear.

Ecosystems are living things, emerging, maturing, dying, and even in death, they add richness to the web of life. Animals have major influences on these systems and the geochemical cycles that humans and all life-forms count on for survival. I can think of no better place to start my exploration of these pathways than on the once-barren rock of Surtsey.

***

I REMOVED THE laces from my trail runners, scrubbed off the caked mud with an old toothbrush, and scraped away any seeds that might've hitched a ride from my home in Vermont or the Icelandic mainland. Only about a dozen people were allowed on the island each year, and in 2021, Borgthór Magnússon, the field leader at Surtsey, had invited me to join their expedition. He'd made it clear that I should remove any stowaways I might be harboring. If a plant or insect wanted a ride to the island, it would have to find a bird or a raft of seaweed.

The day before, I had hiked up the Eldfell volcano on Heimaey, the largest of the Westman Islands, to get a peek at Surtsey. A dark mound about ten miles to the southwest, the island looked like a Devonian tetrapod crawling out of the slate-gray sea.

Bjarni Sigurdsson, one of the lead ecologists at Surtsey, and his grad student Esther Kapinga picked me up at the hotel. We stopped at the local N1 petrol station for supplies. There was a colony of hundreds of kittiwakes on the cliff above the station, flying in to feed their chicks the small fish that they had gathered in the fog. Their poop streaked the rocks like crosswalk stripes.

At the airport, we dropped off our gear as well as the supply of petrol and drinking water for us and the team already on the island. Dressed in black raincoats, Bjarni and Esther identified the plants around an old stone farmhouse near the airport while we waited for the Icelandic coast guard helicopter that would fly us to Surtsey. (It is typical in Iceland to call people by their first names; even the phone book, back when people used such things, was ordered by first name, not last.)

When we returned to meet with the coast guard, the fog over the airport was so thick that the helicopter that was supposed to pick us up couldn't land. The pilot suggested via radio that we drive out to the island's only golf course, where the weather was clearer.

We parked on the edge of the course, which had a beautiful view of the Atlantic, the fog smothering the volcanic cliffs to the north.

"No message yet," said Bjarni as we watched the fog creep over the green.

We got a visual on the helicopter as it approached over two sea stacks known as Haena and Hani, the Hen and the Cock. It circled offshore above the low-lying clouds, then flew away.

As we waited for the pilot to circle back, Bjarni got on the phone. "They're gone."

"Andskotinn!" Esther swore in Icelandic.

It was starting to sink in that we might not get to Surtsey that day—or perhaps at all. We retrieved the water and petrol from the airport and stowed it in the SUV.

Bjarni considered alternatives. We discussed driving to Reykjavík, but the coast guard wouldn't be available. It was a small country, Bjarni explained, and being too pushy wouldn't do the project any favors. (The coast guard ran these trips for the research team as a courtesy.)

When we ate lunch at a harbor restaurant, Bjarni noticed a large rubber inflatable boat tied up below us, mostly used to take tourists out to see whales and puffins. Could that get us to Surtsey? He asked around in case the coast guard fell through entirely but couldn't find the owner.

The fog was beginning to lift by the time we finished lunch, and we could see several fishing boats in the harbor, similar to the ones Bjarni once worked on to pay his way through university. Brawny with blond hair, thick retro glasses, a boyish smile, and a big appetite, he lived up to his name, which meant "bear" in Icelandic. His father had been a fisherman too. "Back in those days, at least two to four of the seventy-five boats in town would capsize every year, with all the crew," he told us. "So you were playing the lottery every time you took a boat. You knew there was at least a five percent chance that you wouldn't make it. It was just part of life."

You were lucky or you were unlucky.

"My father used to always go with the same captain, and one time, for some reason or another, he was too late or something, and he didn't get hired on that boat. So he took a place on another boat that winter. His usual boat went down with all the men. So he was just lucky."

We could see across the harbor to Heimaklettur, or Home Rock, covered in grass and streaked with white seabird dung, a welcome sight to boats returning from sea and those of us hoping to leave. Bjarni called up the coast guard: "If you can come back this afternoon, the ceiling has lifted."

They said they would give it another try. We drove to the airport and paced the empty terminal like expectant parents. The fog poured in like a bad dream, blanketing the wet gray tarmac. We heard the thudding of the helicopter blades, but by then we couldn't even make out the runway from the terminal window.

"They are going to try for the end of the runway," Bjarni reported, still on the phone.

We listened to the muffled blades above the fog. And then it got quiet. The airport attendant gave us the thumbs-down.

"They wished us good luck," Bjarni said. We were on our own.

My heart sank. There would be no second chance for me to visit Surtsey. Bjarni and Esther would lose a year of research, since the biologists on the island were leaving at the end of the week. And things would be tough on the island too. Most of the drinking water and petrol for the five researchers at the station was sitting on the tarmac here.

"I will get to Surtsey even if I have to swim," Esther later said from the back of the SUV.

IF WE COULDN'T be air-dispersed, perhaps we could make it by sea. A crew from the BBC was scheduled to film on the island; maybe we could tag along with them the next day on the inflatable boat they had chartered. Bjarni stopped by the tour company to inquire. The captain asked for a thousand dollars a person. Bjarni wasn't sure if it was in the research team's budget.

We ate a somber dinner and headed back to the hotel. From my window, I saw the sky brighten through the night as the fog lifted. The Eldfell volcano just east of town filled the frame with the black-and-tan abstract shapes of a Clyfford Still painting. On most days, a volcano out the window would be a welcome sight. That evening—which was just one long day—Eldfell taunted me.

The next morning, as Bjarni and Esther did some botanizing, he got a call from the coast guard: "We're out doing fisheries patrols. Can you guys be at the airport in an hour?"

I threw my meticulously cleaned clothes into my duffel bag. We packed up the SUV and drove out to the airport. We boarded the deep blue EC225 Super Puma, all of us wondering if, at last, it was real. Life and travel are a series of lucky breaks and missed opportunities. It seemed that we were finally getting a break.

The pilot tightened the caps on the gas containers and looked up as I brought a big Rollaboard on deck along with the rest of the bags.

"You going to London or something?"

When I stepped off the helicopter onto the tiny, cracked landing pad, it felt a bit like I was walking on the moon—if the moon had a few pioneer grasses, an occasional gull, and a couple of scientists who were older than the ground they were standing on.

Borgthór Magnússon, Surtsey's field leader, made his first trip to the island in 1975 when he was twenty-three years old. Now he had a white, well-trimmed sea captain's beard and wore a tidy zip-up cardigan beneath his Norrona raincoat. As we walked out to the nesting grounds, I asked Borgthór what it felt like the first time he visited. "It was just a heap of ash, gravel, and lava. There were a few plants, and we knew almost all of them as individuals."

Borgthór still knew many of these particular plants, and a tour of the island was a bit like crashing a botanists' cocktail hour. There were a few newcomers, like the coltsfoot and black sedge, and lots of regulars. "I would say the *Honckenya* and *Leymus*"—the sandwort and sea lyme grass, the latter a genus common along the coasts of the North Atlantic—"were the most successful colonizers." These keystone plants spread out over the barren tephra and volcanic sands. Sand dunes formed around the lyme grass. "The largest plant on the island," Borgthór said. "It makes good shelter for the great black-backed gull nests and for their chicks to hide in."

A few of these gulls eyed us from the basalt balconies along the meadow's edge.

There were surprises too. Borgthór pointed out a lone, thick-stemmed plant with long leaves on the edge of the breeding area: the northern green orchid, *Platanthera hyperborea*. "It's quite remarkable for an orchid to grow on Surtsey," Borgthór said, because it needs mycorrhizae, the symbiotic fungi that are essential to providing nutrients to the plant's roots.

Later, as I walked from the intertidal zone to the cabin, I noticed dozens of thin stakes, whittled by the wind, in the lava sand. Many of the island's first plants grew on the eastern edge of Surtsey, where sea and air currents had brought them from the nearby islands. In the first few summers, Borgthór and other researchers recorded every plant on the island. A few of these stakes must have marked the plants Borgthór and his colleagues came to know almost by name. The first were planted in 1968 and used until the 1980s, when they were replaced by GPS. Decades later, the plot looked like a graveyard memorializing the island's first settlers.

ONE MORNING IN the fall of 1963, when Erling Ólafsson was fourteen, he noticed a gray plume rising high above the mountains to the east. As the magma hit the ocean, enormous cauliflower-shaped vapor clouds filled the sky. A dark curtain of ash poured back into the sea.

"I saw the smoke from my bathroom window in Hafnarfjordur," Erling told me. "I stayed for a long time, watching this smoke. Doing nothing, not even moving, just like a mushroom."

Erling never forgot the sensation of watching Surtsey erupt, but he soon turned his attention to something even closer: entomology. When he was young, Erling got a gift from his grandmother—a series called Averdens Dyr (roughly "Animals of the World"), books available only in Danish. Glued to the pictures, he taught himself the language to understand the text. He was drawn to the small invertebrates. By the time he reached university, Erling knew more about Iceland's insects than just about anyone. He caught the eye of a famous Swedish entomologist who was starting a research project on Surtsey.

Erling arrived on the island for the first time in 1970 in a small rubber boat. He set some of those early stakes marking the plant pioneers in the gray lava sands. There was an upper valley, with brown tephra, a flat open plain, a small estuary, and steep cliffs. Surtsey soon came to feel like a younger sibling to him. Earth's newest landmass was still in its geological infancy. "This is the first time that we scientists had land that was quite sterile," he said.

Erling's first trip to Surtsey coincided with the greatest ornithological event in the island's history. A pair of guillemots, small black auks with bright red feet, nested on the island—Surtsey's first breeding seabirds. For these denizens of the Arctic, the newly formed island had a lot to offer: easy access to small fish and krill in the ocean and virgin land free of predators. No arctic foxes. No rats. No people...at least, not until Erling and his colleagues arrived, though they were careful not to disturb the birds of the island's budding ecosystem.

After the guillemots, other marine birds started showing up. Fulmars are classic seabirds, procuring almost all their food from the ocean in the form of small fish, squid, and crustaceans. We could see their eggs, which looked like large golf balls in a lava sand trap, above the cliffs. "The only place in all of Iceland where fulmars actually nest on the ground is Surtsey," Bjarni told me, presumably because predators—and humans—left them alone. Great black-backed gulls feed on fish, birds, and marine invertebrates or scavenge along the coast; lesser black-backed gulls are generalists, usually happy with the insect larvae turned up by a farmer's till or with vegetation, and they and herring gulls are relatively common in Reykjavík.

After the first fulmars and gulls arrived on Surtsey, the great black-backed gulls muscled their way into the prime real estate. The lesser black-backed gulls concealed their nests along the edge of the lava fields. As the colony expanded, so did the green grasses, a ripple effect of feathers, poop, and brawn.

What these birds have in common is that they all bring nutrients to the island. The white streaks of guano around their nests are rich in carbon, phosphorus, and the much-needed nitrogen. In addition to poop, there are carcasses and eggs. During his first visits, Erling recorded each bird he found, creating a timeline of new arrivals. By the mid-1980s, enough gulls and fulmars had arrived to change Surtsey; each of the hundreds of nesting birds released up to three ounces of poop a day—a double shot of nutrient-rich guano.

Near the research cabin, the sharp-edged lava shredded the leather of my pristine boots. When we reached the vast grasses of the seabird colony, it was like entering a different world. Here, the ground felt solid beneath my feet, comforting. There was a faint smell of ammonia and a burst of green so bright, you could see the colony from space, an oasis in the lava sand.

At the edge of the knee-deep meadow grasses, I noticed an increase of *Rumex*, broadleaved sheep sorrels so tall and full-bodied that they almost looked like trees. It was hard to believe there was almost no grass here until twenty or thirty years ago. And there wouldn't be any today if it weren't for the bird shit.

How can you tell that the nitrogen came from the birds and not the atmosphere? The isotopes, or chemical signatures, of the nitrogen in the soil and plants indicated that 90 percent came from seabirds, the rest from the atmosphere. At the center of the breeding grounds, birds deposited up to sixty pounds of nitrogen per acre per year. Outside the seabird area, it was only about a pound per acre per year. (For a rough comparison, farmers typically apply about a hundred pounds of nitrogen per acre of active cropland. Many permanent grasslands, for grazing and hay, receive less than that, maybe twenty-five to fifty pounds per acre.)

Juiced with nitrogen-rich guano, scurvy grass, once a source of vitamin C for sailors, and meadow grass, a native of Europe and Iceland (known in North America as Kentucky bluegrass), began to thrive. Seabird colonies on Surtsey, green with annual meadow grasses, now have thirty times more nutrients and about fifty times more biomass than the black lava fields that surround them because of the guano, eggs, and carcasses from the birds.

These nesting areas are now so lush, the soils so rich, that "we could have cows out there," Erling quipped. "We could have fresh milk every day."

Surtsey sands without seabirds (top) and grasslands in the gull colony (bottom). (Borgthór Magnússon)

As the birds arrived, the plants, and the stakes commemorating them, spread out across the vast lava sands to the southwest. Plants have devised several strategies to spread their seeds. They can fly. They can float. They can stick to the feathers and legs of a bird or pass through its belly and land in its nutrient-rich poop.

As Borgthór and I walked along the edges of the breeding grounds, we were surrounded by lesser black-backed gulls, regarding us through red-ringed eyes, their smooth white heads standing out against the cragged dark lava. It was like being adrift among white-caps. These seabirds were almost entirely constructed of ocean.

A gull called *Tut-tu-gu* overhead, the Icelandic word for "twenty," or at least that's how I heard it. If I couldn't learn Icelandic on this trip, at least I could learn, or approximate, the language of the plants and birds and come to admire the black-backed and herring gulls, often maligned in the city and elsewhere.

Before leaving for Surtsey, I had walked along the Reykjavík waterfront in search of a meal. A herring gull alighted on a café table and grabbed a leftover slice of pizza. Several gulls swooped in. A lesser black-back grabbed the prize. A couple of tourists walked by, and the man stamped his foot at the birds; he and his companion laughed. The gulls flew away. But not without the pizza.

I SAT AT the edge of the seabird meadow watching the fulmars fly in. This was well after dinner, but it wouldn't get dark tonight—or anytime during our stay. I had been lost in the abstract expressionism of the bird-dropping paintings, but here was a Surtsey still life: a gull wing, green sandwort, a white fulmar egg against the dark lava sand. It seemed that the beige palagonite ridge, part volcano, part ocean, had been brooding over the North Atlantic forever, even though it was younger than many of us on the island. Someone walked across the lava field. I could hear it crumble.

A muffled cackle rose from beneath my feet as I walked the grasslands: a fulmar warning me away from its nest. There are about two or three hundred breeding pairs on Surtsey. On the mainland, there are up to two million in summer—more than half the world's fulmars nest in Iceland—but they are restricted to ledges and crevices, hiding away from predators like foxes. *Fulmar* in Old Norse means "foul gull." They feast on the smelly livers of fish and have done well in modern times, enjoying the processed waste and trash from fishing boats.

Young fulmars, once valued for oil and down, protect themselves by stress-vomiting a bright orange streak of grease. Don't get too close. "The spit smells like rotten fish-liver oil, and it's similar in texture," Borgthór said. The distance isn't great, but they can project a few feet. Erling warned me that if I got the vomit on me in Surtsey, I wouldn't be able to rid myself of the smell for the duration of the expedition or even longer. Birds that make the mistake of attacking a fulmar get covered in the foul-smelling goo, rendering them incapable of flight and putting them at risk of drowning.

An adult returned to its nest from the open sea and vomited up a fresh fish meal. Try as it might, the chick didn't get everything. Sloppy eating must bring some nutrients to the burgeoning meadow too. I gave them a wide berth.

Here in the center of the breeding colony, where nutrients are abundant, the number of plant species has declined since Borgthór and colleagues started measuring diversity and productivity in 1990. Outcompeted by the four dominant knee-deep grasses, many of the pioneer plants have disappeared. There is now a dense grassland where ten different species once grew.

Things get interesting along the edges. The ecological processes are at their most dynamic at the border between the chaos of lava and the spongy wealth of grass—at the boundary of the ripple, where the new birds, mostly lesser black-backed and herring gulls, are moving in, helping to gentrify Surtsey, transforming it from a pioneer community to a grassland. Seventy-eight species of plants have been identified on the island. The lowest diversity was on the lava field, no surprise, but the number of species—if not their abundance—is also relatively low within the bird colony, where the nutrients are high.

Only a few grasses dominate. The lushest parts of the grassland are practically monocultures compared with the fringes, where grasses give way to barren lava.

This borderland reminded me of what ecologists call "the intermediate disturbance hypothesis," developed to describe trees in rainforests and animals that live in intertidal zones. Stable areas—like the meadows of Surtsey, where a few dominant grasses with plenty of nutrients outcompete other plants—allow a few species to thrive. In contrast, ecosystems under constant change are tough places for animals and plants. The rocky shoreline at Surtsey, where new lava boulders tumble down continuously, is too dynamic for many species to take hold. Moss doesn't grow on a rolling boulder. The sweet spot for biodiversity, where new plant species can find a niche, is often the intermediate zone: not so many nutrients or so much stability that a few species take over but still stable enough that an emerging community won't be wiped out entirely by a sudden change in the landscape, and even newcomers can survive.

It brought to mind the Icelandic Viking origin story: In the beginning was chaos—in the north, snow and ice; in the south, heat and fire. Life emerged on the land between the two.

A FEW YEARS ago, several scientists on a research cruise in Baffin Bay, a pristine part of the Canadian Arctic, were surprised to come across an area with high concentrations of ammonia, something you might expect to find along an industrialized, polluted coastline. The models that atmospheric scientists like Jeff Pierce of Colorado State University had put together suggested that there shouldn't be any ammonia in that remote part of the Arctic.

And then the scientists looked out the window, or perhaps just at a chart. "The ammonia concentrations were highest when the ship was near places known to have seabird colonies in the summertime," Pierce told me. That made sense; seabird poop in big colonies often emits gases that are rich in nitrogen. He and his colleagues added an inventory of migratory seabirds to the model. "We realized that seabirds were almost certainly the missing source of ammonia in the Arctic."

This pungent gas can hook up with sulfuric acid, abundant in the region, to form particles. The particles form droplets. Clouds with more droplets are denser, and they appear whiter and brighter. Pierce likened it to looking down on a glass of water on a black table. "If you put three ice cubes in, there's going to be some light reflected by those ice cubes, but for the most part, you'll see the black surface. Now, if you took those ice cubes and crushed them into tiny ice fragments, they would be really good at reflecting the light from above." So if you looked down at the glass with the crushed ice, it would appear white, even though it had the same amount of ice. The ammonia from the seabird colonies formed lots of small droplets. The clouds still held the same amount of water, but like the crushed ice, they now had a lot more surface area and reflected more sunlight back into space.

"So you have this effect on climate," Pierce said. The clouds over seabird colonies keep the Earth cooler because they're brighter, with the biggest effect in areas with the most birds. Large colonies are found from the Arctic Archipelago, north of Canada's mainland, to Iceland; the diffusion of ammonia can extend hundreds of miles from the seabird colonies. Birds help keep the Arctic a little colder, perhaps in their own small way dampening the effects of climate change one splat at a time.

ONE OR TWO new plant species have been recorded on Surtsey in each year of its existence. The first ones came by sea, some others by air, taking root as the nitrogen built up on the island. But the vast majority, about three-quarters of the seventy or so established plants, came on the wings, guts, feathers, and legs of birds, mostly gulls.

Insects, some arriving on the winds, others on the wings of birds, started settling in. More than three hundred species of beetles and other terrestrial invertebrates have been found on the island, including a weevil so rare that it was thought to be new to the planet until others were found off the coast of Scotland. The insects are painstakingly collected with tweezers, paintbrushes, or straws or by dragging white sheets over the grasses. At least 143 species are considered permanent settlers. Regardless of their status, each new species is a cause for a celebration.

Over time, the insects attracted insect-eating birds: snow buntings, meadow pipits, and wagtails. A few graylag geese flew in from the Icelandic interior. The gulls were not impressed. They screamed and honked at each other. "It is definitely not a happy marriage between gulls and geese," one ornithologist told me.

Every year, the fabric of the island has gotten thicker, more lustrous, and more diverse. Gray seals showed up in the 1980s. They hauled themselves out along the northern spit of the island, gave birth, and nursed their white-haired pups. The pups pooped. The adults pooped. Add to that the placentas and the occasional carcass, and you've got a marine subsidy, a nutrient supply moving from the ocean to the land. It is smaller than the seabirds'—about twelve pounds of nitrogen per acre—but in a new area with greater access to the sea. For the island's avian scavengers and plants, these nutrients have been a reliable and substantial yearly bonus. Seals like coastal flats, not cliffs, so their breeding opened the lower shores to oysterleaf, sea rocket, and saltbush. They created a seal oasis among the dark gray sands and lava boulders of the north.

Surtsey's seals, like its seabirds, demonstrate how animals can move nutrients onto barren land and, in so doing, create full-fledged ecosystems from scratch. Sable Island, a strip of sand more than a hundred miles off the coast of Nova Scotia, provides another example among many. In the past fifty years, the number of gray seals that pup on the island has grown from a few thousand to more than ninety thousand, making it the largest gray seal breeding colony in the world. The nitrogen the seals bring fertilizes the dune grasses of the island, now home to a population of feral horses—living proof that Erling's herd of dream cows on Surtsey's seabird meadow is possible.

The plume of nitrogen from the seal poop stretches out into the waters beyond Sable Island, increasing the phytoplankton on its leeward side by 20 percent. Though the algae are microscopic, you can see the seals' signature—a bright green splash of chlorophyll—from space, like that of the seabirds on Surtsey.

"Whenever people of my generation have a nightmare, we're dreaming of running away from a volcano, right?" said Freydís Vigfúsdóttir, a seabird biologist who was born on the Westman Islands in the 1980s, well after Surtsey cooled. "I've walked on many lava fields before, but the one on Surtsey is different. I could hear rocks falling underneath me. The sound of waves. Obviously, there were caves beneath us, and I remember thinking, *If I fall, no one's ever going to find me.*"

On Surtsey and on other, more accessible parts of Iceland, such as the Snæfellsnes Peninsula, I've occasionally been touched by a sense of awe. The barren landscape—carbon-black and brick-red mountains cloaked in gale-force winds beneath a rising moon—brought to mind the Romantic sense of the sublime. There was beauty in it. But there was terror too.

"If nature decides this is the time," Freydís said when we chatted a few months before I traveled to Surtsey, "then you're toast." She shrugged.

Most Icelanders had no interest in letting nature take its course. "Lava was seen as something very, very ugly," Bjarni told me in Heimaey. "It's really psychological. If the locals could control the lava fields, they would flatten them and put in a lawn." Or at least put in some lupine, a purple flowering plant introduced from Alaska that can supply its own nitrogen via symbionts in its roots.

During their stay on Surtsey, Bjarni and Esther focused on the microscopic and microbial differences in the soil across Surtsey, ranging from the vegetated bird-breeding areas to pure pumice. I followed them to one of their sampling sites. Bjarni unearthed two-year-old tea bags from the ground. "I ask the microbes, 'Do you want red or green tea?'" The TBI—tea bag index—is used around the world. To keep things consistent, they use only Lipton. "We're all nervous," he told me, "that Lipton will stop making red tea."

Esther took a core sample through a dead bird.

While they were running transects earlier that day, Bjarni had noticed a ringed plover giving him and Esther the eye. When he looked straight at it, the bird flitted away, part running, part flying, staying low to the ground but clearly visible. Plovers had been seen on the island before, but no one had ever found a nest.

"When you see a plover running or flying low to the ground, always go the opposite way to find its nest," Bjarni said. The ringed plover is a humble bird, its nest little more than a scrape in the volcanic sand. Bjarni ignored the bird's attempt to lure him away and found three speckled eggs on top of the lava. A first for the island, and the seventeenth species of breeding bird.

THE ABSENCE OF humans, it turns out, requires careful curation. In 1969, one of Surtsey's researchers found a leggy plant with sawlike leaves that stood out among the sea rockets and grasses. It looked like a new species to the island, so he called in an expert, who turned over the rocks and found some unusually rich soil amid the lava. It was *Solanum lycopersicum*. A tomato plant had rooted in a visitor's night soil. The plant and the poop were bagged up and carted away.

To avoid the same mistake, we drop our drawers at the edge of the waves. With high ceilings—after the fog lifted—a long distance between the walls, and a view of the Westman Islands, chalk white with gannet poop, it's the most magnificent bathroom in all of Iceland. You can even get a flush, but there's a catch. You have to perch on the round lava boulders and time it right, preferably when the tide is low, so all of your poop will be washed out to sea. The ocean can make a ferocious noise, tumbling the boulders before tossing them ashore.

Volcano ecologist Charlie Crisafulli thinks of colonization after eruptions as musical chairs. Luck and timing play a big role in what persists in the volcanoscape on the mainland. If a stand of trees or group of animals escapes the lava, they can disperse their seeds or move out from their refuge when the eruption ends. A grasshopper or beetle on the edge of the eruption can hop, fly, or crawl in.

Surtsey, though, is more like ocean roulette. "It's a very small target in a big frigid sea, and most of the adjacent land is already depauperate," Crisafulli noted. The Icelandic mainland is isolated along the Mid-Atlantic Ridge, with only a few animals and plants. It has no native terrestrial mammals except for the arctic fox (which crossed the sea ice in the Little Ice Age); it has no amphibians, no reptiles, and no mosquitoes. There are only a few animals anywhere near Surtsey, and those that make it to the island find it a hard place to survive. There's no standing fresh water, so for ducks and waders, it is an inhospitable place to breed. Plus, it's cold, windy, and foggy.

We had learned this the hard way on the tarmac in Heimaey. We arrived here in part because of Bjarni's determination and the generosity of Iceland's coast guard. But mostly, it was luck—a small weather window when the helicopter happened to be in the area. The BBC filmmakers tried just as hard to reach the island, but when they showed up in their boat off the northern spit, the wind, waves, and boulders made landing too treacherous. Years earlier, Bjarni noted, a photo from Surtsey had been chosen as one of Iceland's photos of the year: "It was a picture of my PhD student falling out of the boat when we were landing."

The BBC film crew had to turn back, and that could just as easily have been our fate. For all the birds, plants, and insects that reached Surtsey, many others did not; they died, took another turn, or never left the comforts of home. For all of us, it could have gone either way.

Not long after Surtsey emerged, Iceland's botanists, geologists, and ornithologists banded together to keep it pristine, limiting

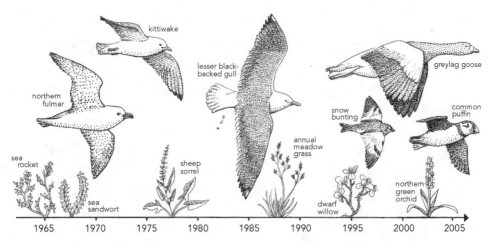

A Surtsey timeline. The earliest plants to establish themselves on the island were sea-borne species; they packed their nutrients for the journey. After the arrival of the lesser black-backed gull and other seabirds, nitrogen from guano helped catalyze the establishment of meadow grasses. The future might belong to the puffin as the island erodes away. (Based on Magnússon et al., 2009)

annual visitors to a few researchers. Even at full occupancy during the expedition, the island was home to only a handful of scientists. But for more than three hundred and fifty days a year, Surtsey is completely free of humans. The birds, insects, and plants are the true owners of this island. And even the birds abandon the island in the winter. With fewer than six hours of sunlight between November and January, darkness consumes Surtsey. The plants hold down the island in the gloom with no visitors other than the occasional marine mammal swimming by.

There are plenty of rocks in the ocean, but Surtsey is one of the youngest, with the lightest of human footprints. "Surtsey is the only volcanic island that has been studied closely from the very first day," Erling said. In 2008, it was added to the UNESCO World Heritage List because of this protection and the biological studies conducted there.

There have been plenty of eruptions on the mainland, but there are animals and plants around those lava fields or relict communities that survive the blasts to help seed recolonization. "Surtsey taught us that seabirds could bring nutrients, seeds, and insects in nesting material," Erling said. There are so many invertebrate species on the island that the source can only be nesting birds.

As I sat on the edge of the meadow, I thought of another volcanic island half a world away, described by a different scientist who had first landed there when he was a young man in his twenties. The Galápagos, "a broken field of black basaltic lava, thrown into the most rugged waves," brought Darwin close to "that mystery of mysteries—the first appearance of new beings on this earth." At least a few of the beings he observed probably got their start on islands enriched by seabird dung.

"I'll never forget the feeling when I left Surtsey after the first summer," Erling told me later. He'd returned to Heimaey, with its own history of eruptions and colonization. The grassy hills were unsettling. "I had not seen a green color for three months," he said. Even after Erling returned to Reykjavík, Surtsey was never far away. The brown hills and gray sands became the palette of his life. "It is the home of my heart. As simple as that."

PÁLSBÆR, THE RESEARCH cabin named for Paul Bauer, an American philanthropist, rides the edge of the lava field like a fishing boat on the dark winter sea of the North Atlantic. The researchers gathered around the wooden table at the heart of the cabin in the long dusk that passed for evening on Surtsey in July, bathed in light from two candles. The Westman Islands could be seen out the window. One chalky block, flecked with gannets and covered in guano, was as bright as a lighthouse beacon.

There was a celebratory air. Earlier, Borgthór had been walking along the edge of the breeding colony, and when he came back to Pálsbær, he casually mentioned, "One hour ago, I found a new species on the island." It was a blue sedge growing between the grassland and lava field. *Carex flacca* is common in Iceland, but here it was a big discovery. Likely bird-dispersed, perhaps by geese, it had probably been on Surtsey for a few years. It was the seventy-ninth plant species found on the island, evidence that its diversity was still growing, thanks to the birds.

Borgthór poured some wine into a brown mug—the honorary Sturla cup—and handed it to me. (Sturla Fridriksson wrote several books about Surtsey, including the one I had used for reference before arriving.) Icelandic lamb was served. The researchers had spent the day counting—plants, birds, a few thin strips of seaweed—and the talk inevitably came around to the culinary properties of seabirds.

Although the population of Iceland is smaller than 400,000, culinary traditions vary across the island. Fulmar chicks are favored by some; young black-backed gulls by others. Gannet chicks are collected on the steep white cliffs we could see out the window. Soup made of puffins, with their heads left on, was a favorite in Heimaey, where locals capture the birds in nets around the nesting areas. In the fjords of the east, shags, also known as cormorants, are roasted, salted, or smoked. "They have delicious brown meat," noted one of the researchers. Perhaps it wasn't the late-night one-upmanship of marine biologist versus fisherman on the *Orca* (Quint's boat in *Jaws*), but there was an air of camaraderie and competition in the unyielding twilight.

The birds could fight back. A great skua, a fine pirate of a bird, once attacked Bjarni's uncle. "He was a big man but nothing against this Messerschmitt. The skua knocked him cold."

Borgthór admired fulmars not for their culinary properties but as survivors. To my untrained eye, they looked like gulls, but they are closely related to giant albatrosses. And they have good genes. "There's a picture of my mentor, George Dunnet," said Borgthór, "tagging a fulmar when he was a young man. Fulmars don't breed until they are ten years old, and then they have one chick per year." He showed me two black-and-white photos. The one on the left, a fulmar and a man with dark hair and smooth skin, was taken in 1951; the one on the right, a fulmar and an older man with a receding hairline and facial grooves, was from 1986.

"It's the same person and the same bird," Borgthór said. The fulmar hadn't aged a day.

ON THE LAST day of the expedition, everyone was up by six checking insect traps, finishing soil samples, flying a drone over the northern spit. We all mustered at eight to survey bird nests. Borgthór used a sixty-foot rope to define a thousand-square-meter circle. Holding the rope tight, we walked in a circle around Borgthór, starting at the center of the breeding colony.

"Hreidur!" someone called out, the Icelandic word for "nest." In the first circle, there was only one gull's nest, a ring of dead grass among the densest lyme grass and chickweed. After twenty years of hosting seabirds, the ground was soft and spongy, disguising the gaps in the lava below. It was a bit like walking on a trampoline, with the occasional white feather drifting through the waist-high grass.

We did our best to count the birds and minimize disturbance. I heard a croak beneath my feet—a fulmar ready to projectile-barf a stinking gob of fish guts. I gave it a wide berth. The gulls had retreated, their nests little more than a cowlick: a small bed of straw, a few feathers, perhaps the remnants of an egg. An embedded naturalist, I did my best to hold on to the rope through the meadow. The crooked lava hidden beneath the grass felt like a sea frozen in mid-gale.

"Hreidur!"

We recorded thirty-four nests that day, many of them on the edge of the grassland, helping the habitat expand over time.

THE RESEARCHERS, TOO, were dispersed by wings, or in this case rotor blades. As we waited for the helicopter, I mentioned I was thinking about booking a hotel room. The photographer leaned in. "I would wait." This was Iceland, and it was too soon to plan. The helicopter was still about twenty minutes away.

We watched it approach over the horizon, touch down on the tiny landing pad, and unload the geologists who would take our place at the hut. We put on our headphones, loaded up our gear, and were off. I had no idea where we were going. Back to Heimaey? To Reykjavík? Iceland's unofficial motto is *Thetta reddast*, pronounced "*Thay*-ta *ray*-dast" and roughly translating to "It'll all work out." As we lifted up over the North Atlantic, I felt like a hitchhiker, little more than a burr in the belly of the whirlybird.

I was on the island for maybe seventy-two hours all told, though it felt sort of endless and unpunctuated, since it never got dark. Surtsey disappeared. One of the coasties passed back a note: *Reykjavík at 18:15*.

WHEN I FIRST spoke to Erling in 2019, he had just celebrated his seventieth birthday; he was a man shaped by the seabirds and seals of the North Atlantic just like Surtsey itself. Seventy is the hard stop between work and retirement in Iceland. But even on his last trip, Erling told me, his sparse gray hair emerging as if from fumaroles in the breeze, "there is always the same excitement. Every time. I just leave the boat or helicopter and the first thing I do is grab a handful of sand and kiss it."

At the end of Erling's last trip, in 2020, Surtsey wasn't ready to let him go. The departure of the research crew was delayed for one day, then several more, because of bad weather. The coast guard helicopter couldn't land on the tiny landing pad, and a boat was out of the question. Supplies were running low, and the next team was overdue. In the end, the coasties dropped down a rope, and Erling and crew were plucked off the island like wildflowers.

When I visited him in his office, in a building on the edge of a parking lot and a lava field, Erling was surrounded by the thousands of beetles, flies, and spiders, pinned and mounted in boxes, that he had collected in Iceland over the decades. He was organizing his life's work for the next generation of entomologists, and he looked a bit like a man in exile.

He still seemed heavyhearted about leaving Surtsey. "It was like being torn up by my throat," Erling said with tears in his eyes. "Erling was almost born on this island," Borgthór said. "Or at least, he grew up here." Erling told me wistfully that he had considered joining the Surtsey expedition this year but couldn't get through the pain of saying goodbye yet again.

I stopped in the bathroom on the way out. Above the toilet, there was a picture of a cormorant defecating on the shore. It was one of Erling's most prized photos.

EAT, POOP, REPEAT. The guano effect ripples far beyond Surtsey. Seabirds breed in the Arctic, in the Antarctic, and on islands throughout the world. The Southern Ocean is home to the greatest number of seabirds—think penguins, petrels, and albatrosses—and about four-fifths of all the world's seabird poop and the nitrogen and phosphorus it contains. Surtsey's story has played out on islands around the world, in some cases for centuries, even millennia. Guano is a precious natural resource, and its high concentrations in the Southern Hemisphere once prompted a global chase, from Peru to the guano islands of the South Pacific.

In the nineteenth century, sailing vessels from Europe and North America stopped at remote islands around the world. It was a boom time for whale oil and bird poop. The oil was used for light and lubrication in the big cities of the north. The guano supplied nutrient-depleted fields and croplands with nitrogen and phosphorus—it was considered the best fertilizer in the world, as we'll discuss later. Guano extraction imperiled many seabirds by destroying their island burrows and nesting habitats. The harvesters often persecuted native seabird predators, such as Andean condors and peregrine falcons, and their commutes between harbors and breeding islands also shuttled in invasive predators. Perhaps none has been more damaging than *Rattus rattus,* the ship rat.

"It's chalk and cheese," Nick Graham of Lancaster University said of the difference between a rat island and a rat-free one. He was waiting out the coronavirus pandemic with his three offspring in the United Kingdom when we spoke. Graham has worked on the Chagos Archipelago in the Indian Ocean for more than a decade. The islands he studies are more or less identical except that some have a history of shipwrecks and fleeing rats. For the native wildlife, it was yet another game of ocean roulette, and the arrival of a new predator was like a bullet to the head. On the rat islands, the rodents ate seabird eggs, chicks, and even adult birds on occasion. This had ripple effects throughout the ecosystem.

"When you set foot on an island without rats, the skies are full of seabirds. It's noisy because of the cacophony that those birds are making. And it smells of guano and ammonia, particularly if it has recently rained. It's a really rich, pungent, loud environment.

"But when you set foot on an island with rats present," Graham said, "there's next to no seabirds. The skies are empty." There is no smell, and the only sound comes from the small waves lapping on the beach.

Graham wondered how these differences might affect the islands' reefs and vegetation, so he took a risk and spent the last of the grant money he had from the Australian Research Council. On six islands with rats and six without, he and his colleagues collected soil samples and new leaf growth from a coastal shrub, then they snorkeled out to the reef flats and gathered macroalgae and solitary sponges. From the start, the difference in the number of seabirds was obvious: there were seven hundred fifty times more seabirds on islands without rats than on those with them. More birds meant more poop. The deposition of nitrogen was two hundred fifty times higher on the islands with birds, an enormous resource for the native plants and animals.

They dived off the reef crest, where the corals drop down to deeper water, and counted fish and collected turf algae and jewel damselfish, which feed on seaweed. Damselfish grow at a faster rate on the reefs of islands without rats. Known as the gardeners of the reef, they defend their algal farms, protecting the small shrimp and the nutrients that their poops provide for the algae. Fish biomass is 50 percent higher on the reefs with seabirds than on those without, which was surprising, considering the reefs were unfished and already in good shape. Graham and his colleagues are now interested in the impact of seabirds on fish fecundity. With more nutrients, the fish could have more offspring dispersing out to nearby islands, expanding the footprint, the poop-print, of the birds.

But can't too many nutrients be bad for corals? We've seen declines around cities without good sewage treatment. "People often think of nutrients as being bad news for coral reefs," Graham noted. That's because a lot of human inputs are from fertilizers and sewage, which are phosphorus-limited and have lots of nitrogen. If the corals are impaired by this excess nitrogen, they will bleach at a lower temperature, kicking out their essential symbionts. "But if you increase nutrients with a balanced input of nitrogen and phosphorus, which is what the seabird guano provides, corals grow faster," he said. Corals will stay healthy at higher temperatures, retaining their symbionts, and be more thermally resilient in the face of climate change.

It's no surprise that the birds and their poop, eggs, and carcasses enhance the growth of plants—Surtsey and other studies showed that years ago—but marine fishes respond to this subsidy too.

AFTER THE BIRD survey, Borgthór and I sat on the edge of Surtsey's vertiginous southern cliff. We could hear the island falling apart below. I had assumed that the round lava boulders I stumbled over while walking along the shore were a couple of decades old, but Borgthór said that many had likely fallen into the water mere months ago and were tumbled and rounded by the surf.

I spent a long time staring into the white splats on the dark lava during that endless summer afternoon. When I looked up, the seabirds returning to the cliffs stitched the sky like a dark seam on a baseball. After feeding offshore on capelin, sand lance, and krill, they carried some back in their bills or gullets for their chicks. A fulmar, likely with a taste for fishing-boat discards, breezed by overhead. A puffin circled the cliff edge. Black guillemots returned from the sea, wings flapping quickly before they swooped into their nests, the white stripes of feces their landing strips. Calling overhead, the black-backed gulls made their way to the meadow they had formed. Though they're just gulls elsewhere, perhaps fighting over a piece of pizza, they will always be glorious birds on Surtsey.

There was a tug-of-war between the physical forces that had built this island, and were now inexorably eroding it away, and the biological ones, the nutrients and the accreting biomass in the foothills of the barren palagonite. Many of Surtsey's seabirds are older than the boulders, and some of the fulmars might even be older than the island itself. Here was a rare opportunity to observe how animals could build an ecosystem almost from scratch. You could see the process unfolding during a short afternoon hike, or, if you were a biologist, you could follow it over the course of a career. Any scientist will tell you that research projects often have high turnover, with young people coming and going as they move into academic, government, and nonprofit careers, but the team studying Surtsey has had excellent retention. Many scientists, like Borgthór and Erling, have made annual visits over the course of their careers. This would be Borgthór's last expedition. He was retiring later that year.

Nothing lasts forever. Several of the smaller islands to the east emerged about five thousand years ago. They've been reduced by erosion to basaltic shards, steep cliffs with gannet nests and species-poor grasslands. They portend Surtsey's future. And Surtsey shows us what the old seabird outcrops looked like in their youth: supple, constantly changing.

At about three hundred acres, Surtsey has already shrunk by about half in the five decades since it was formed. Its profile is a little more chiseled, with steeper cliffs and the northern spit now sticking out like an elf's hat.

"In the future," Borgthór told me, "this will be the land of the puffin." The adorable bright-beaked seabirds will burrow beneath the thick grasses, as they do on neighboring, steep-cliffed islands. (More than half the world's Atlantic puffins live in Iceland.) The fragile lava will slough into the sea, leaving behind a hard inner core of palagonite, the basaltic glass that formed when the lava, still hot, flowed into seawater. "Eventually, maybe in ten or fifteen thousand years," he said, "Surtsey will probably be gone."

He let that sink in.

"But then we will have another eruption and a new Surtsey."

2

Deep Doo-Doo

In the great expanse of the oceans, animals don't seem to matter very much. From the surface, it all looks pretty much the same. A lot depends on the wind and the currents—the physical forces at play. The phytoplankton, microscopic plants, are the showrunners. Tiny in size but numbering in the trillions per square mile, they form the base of the food web, taking in sunlight, carbon dioxide, and other nutrients and converting them into seasonal blooms that feed zooplankton, fish, and, ultimately, the largest animals that have ever existed. If you're ever lucky enough to see one of these whales up close, to hear her blow, smell her breath, and taste the brine, your perspective on the emptiness of the ocean might start to change. What if these enormous mammals, through their poop and their pee, affected one of the main drivers of photosynthesis—and, thus, productivity—on the planet, much as seabirds jump-started plant life on Surtsey?

In the nineties, I volunteered with the Right Whale Research Project, based at the New England Aquarium, working to save one of the world's most endangered whales. We traveled east from Maine to the Bay of Fundy in the Canadian Maritimes aboard the *Nereid*, the aquarium's twenty-nine-foot research vessel. I was put on watch and told that these filter-feeding whales "log," or rest at the surface, after long deep dives to the muddy bottom of the bay. As we sailed past Grand Manan Island, I saw a shadow to the east. We stopped, lifted our binos, and glassed the jeweled surface of the bay. This logging whale turned out to be, well, a log—the wooden kind—most likely a dead spruce taken by the tides.

Not long after that false alarm, we saw a faint mist in the distance, like a speck of dust on the horizon. As we got closer, I could see the V-shaped blows and dark scowl typical of a right whale—the goths of marine mammals. One of the researchers recognized the whale as 1227, an adult male (he later went on to sire three calves and is still regularly sighted around the Maritimes). Sliver, as he was known, broke the surface, took several breaths, then lifted his broad black flukes. Just before he dived back under, 1227 released an enormous brick-red slick.

"Poop!" someone yelled.

As if we needed to be told. An overwhelming stench of brine and decay drifted our way. That fecal plume marked the beginning of a new course for my life. I had no idea at that point that I'd still be following whales around and collecting their feces more than twenty years later.

WHALES HAVE BEEN hunted commercially for at least a thousand years. Many populations declined, starting in Europe and then spreading out around the world, as whalers expanded their range in search of baleen and oil. By the early twentieth century, whalers had explosive harpoons, diesel chaser boats, and factory ships that could render entire blue whales, the largest animals on the planet, at sea, and no whale was safe from industrial whaling. By the 1960s, many species were on the brink of extinction. The

North Atlantic right whale population was probably down to fewer than a hundred individuals. Ninety-nine out of every hundred blue whales had been removed from the vast whaling grounds of the Southern Ocean.

Protections like the U.S. Marine Mammal Protection Act and the moratorium on commercial whaling imposed by the International Whaling Commission helped turn these trends around in the 1970s and 1980s. The return of whales in large numbers since their near extinction has been a cause of celebration in much of the world, but there has also been some backlash. Starting in the 1990s, Japan and other whaling nations came out with strong statements in defense of the commercial hunt. In addition to the cultural claim that whaling was important to Japanese heritage and traditions, the nation had two other lines of defense: First, whales ate a lot of fish, and too many whales had a negative impact on fishing communities. Second, whales were bad for conservation. By the 1990s, the diminutive minke whales, the focus of hunts in Japan, Norway, and Iceland, were so numerous that they were outcompeting the larger endangered whales, like fins and blues—or so the Japanese claimed. By killing the common minke, they contended, whalers reduced the competition for more endangered whales.

This perspective, looking at whales and other species simply as consumers, was not uncommon at the time. Much of marine mammal ecology at that point focused on eating—the consumptive effects that species had on each other. In 1997, when I was taking marine ecology as a master's student at the University of Florida, the "Whales eat our fish" argument might have been in the back of my mind, since I had recently gone to Japan with researchers who were examining commercially sold whale products using genetic techniques. They had found that some whale meat sold in markets and sushi restaurants throughout the country came from endangered species, a violation of international law, and it got me wondering if there were other misconceptions guiding our management of whales and other ocean creatures.

So I was probably daydreaming in the back of the room when Larry McEdward, a larval ecologist and professor of zoology, drew the biological pump on the blackboard. This is a basic concept in biological oceanography, the premise of which is that carbon and other elements move from the surface to the deep sea. At the ocean surface, there is light—and thus photosynthesis—as well as an interchange of carbon dioxide from the atmosphere. When phytoplankton, zooplankton, and fish die, they transport carbon and other nutrients to the bottom of the ocean when they sink. This transport also occurs during the vertical migration of zooplankton.

The movement of zooplankton is one of the great, mostly unseen migrations on the planet. Krill and copepods typically feed on phytoplankton at the surface at night, then migrate back to the depths of the ocean by day, presumably to escape predators by hiding in the dark. This daily vertical migration forms one of the largest animal movements on Earth. Their deep-water defecations and deaths transfer billions of tons of carbon away from the ocean's surface every year. The surface waters—the only place where there's enough light for photosynthesis to occur—can be depleted of nutrients, but they build up in the deep sea. This stratified system of warm, nutrient-poor surface waters and cold, nutrient-rich deep waters generally occurs in the summer, when there's lots of sun and little wind to mix things up. The nutrients at the surface decline just as they do in your garden after a productive season. In some coastal areas, winds and upwelling can bring the nutrients back up in the fall.

Meanwhile, marine snow—tiny biological particles made up of fecal pellets, dead phytoplankton, and other debris—also sinks into the depths. The combination of these two processes moves nutrients to the aphotic zone, which has too little light for photosynthesis. This biological pump plays an important role in the export and storage of carbon, nitrogen, phosphorus, and iron to the deep sea and ocean bottom.

But there was something big missing in McEdward's presentation that afternoon. I remembered those right whales surfacing with mud on their bonnets after feeding at depth for ten or fifteen minutes. Many animals travel up and down once a day, but deep-diving whales and other air-breathing vertebrates can journey from the water's surface to the depths of the ocean and back multiple times per hour. Dives are energetically expensive, so these animals shut down many basic metabolic processes while they feed. Whales return to the surface to breathe, of course, and also to rest and digest. And sometimes, just before they dive again, they leave behind enormous fecal plumes. This was the opposite of the biological pump, I thought; in this case, animals were bringing nutrients to the surface rather than exporting them to the deep sea.

I scratched a diagram in my notebook reflecting what I had seen at sea over the previous years—right whales eating copepods at the bottom of the bay, swimming to the surface to breathe and poop—and what I assumed happened: the nutrients in the poop were picked up by the phytoplankton, which ultimately fed the copepods and the whales. I forgot about this drawing and lost track of it over a series of moves, but the concept stayed with me. Then one summer, when our family was insulating our attic and I had to remove things I'd stored there, I came across my notebook from marine ecology in an old liquor box. It opened to the original sketch.

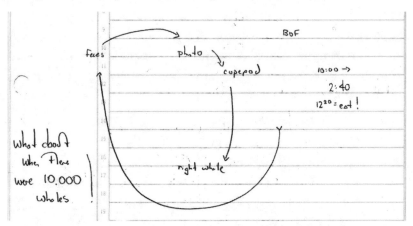

The whale pump as sketched out in my marine ecology notebook during a class at the University of Florida in 1997. BOF: Bay of Fundy. (Note that as a graduate student, I had to remind myself to eat.)

I had discussed the idea with colleagues at the University of Florida, but it wasn't until a dinner conversation a few years later with Jim McCarthy, a close friend and mentor at Harvard, that I found an oceanographer willing to entertain the idea that large animals might be ecologically important in the oceans. You might think, *Well, of course whales matter, they're big and cosmopolitan, found in every sea.* But that isn't how the field of oceanography had developed in the twentieth century. Most of the biological oceanographers I knew focused on bottom-up forces: the relationships between nutrients, ocean currents, phytoplankton, and zooplankton. Jim was open-minded and an expert in the marine nitrogen cycle. He had the skills to measure these nutrients.

To understand the implications of this work, it helps to sketch out some basic features of ocean ecosystems. In the Gulf of Maine and much of the North Atlantic, the ocean is well mixed in the winter, meaning it has lots of nutrients but the wind and reduced sunlight keep the growth of phytoplankton, microscopic plants, limited. Then in the spring, the waters warm and a stratified layer, or thermocline, forms; this keeps the phytoplankton near the surface, where there is plenty of light for growth and lots of nutrients from winter mixing. Phytoplankton become so abundant they create a spring bloom. Zooplankton, such as copepods and krill, respond by feeding at the surface. Herring, sand lance, and other small fish eat the zooplankton. Many of these animals hide in the deep aphotic zone at night. With all of this potential food, large predators like tuna, seabirds, seals, and whales migrate into the area.

Here's where the whales come in. As they dive to feed and then surface to breathe, rest, and digest, they move nitrogen and other limiting nutrients—fertilizer—across the thermocline. Nutrients that were locked away in the deep, dark cold waters are moved to the light, warm surface waters, where they can be used for photosynthesis. Whales can do this physically, by swimming through that barrier, and also by moving nutrients, via feces and urine, from the depths to the surface. This movement becomes even more important over the summer, when the nutrients near the surface begin to run out; whales are fertilizing their feeding areas.

Whales, seabirds, dolphins, and seals—animals that feed in the ocean and are tethered to the surface to breathe—all play a role in recycling nitrogen in the Gulf of Maine, but whales, with their vast size, have the largest impact of all—a finding that is all the more remarkable when you consider that present-day whale populations are a fraction of what they were a few hundred years ago, before the advent of industrial whaling. We estimated that whales moved twenty-four thousand tons of nitrogen to surface waters each year in the Gulf of Maine—that's more than all the rivers combined, a natural source of the nutrient for the region.

A colleague suggested a term for this process: the whale pump. McCarthy and I submitted our first whale-pump article, based on a model we had created, to a leading journal in the late aughts. It was rejected. There was pushback from some in the oceanographic community, which focused on bottom-up processes. Some argued that it was the small players—bacteria and phytoplankton—that drove marine ecosystems along with physical processes, like upwellings and storms. What difference could a few whales make? One of the reviewers who was willing to entertain the idea that this movement might be an important, overlooked process noted that we had a bunch of whales swimming along the New England coast, not far from where we were based. Why not get out in the field, collect samples, and measure the direct effect of fecal plumes on the system?

"Start again" was the last thing we wanted to hear, but colleagues were putting suction-cup tags on humpbacks that year to follow their movements beneath the sea, so in June, I joined them on Stellwagen Bank, off the coast of Massachusetts. We collected as many fecal plumes from humpback whales as we could, along with samples from the occasional fin and right whale.

Popular Science once called "whale-feces researcher" one of the worst jobs in science, but I didn't see it that way. Whale fecal plumes can be neon green or bright red. There are floating bricks and leather-brown stools. At times, they sparkle with scales, like the sun glinting on the water. Every whale defecation is unique.

"I found that snowflakes were miracles of beauty," Wilson Bentley wrote after photographing thousands of crystals at his home in Jericho, Vermont. "Every crystal was a masterpiece of design, and no one design was ever repeated." (The meteorologist later earned the nickname Snowflake Bentley for his work.) It might be pushing it to say that a whale's fecal plume has the symmetry and beauty of a snow crystal, but no two poops are alike. When whales are feeding on lipid-rich crustaceans, their poops tend to float and clump into bright red, flocculant feces. (Lipids are carbon- and energy-rich molecules; common forms are fat and blubber.) When they feed on fish, their poops can be more subtle, a cloud of unknowing, like oversteeped green tea. One afternoon, we noticed a huge vertical spout off the stern of our Zodiac. The sound was as deep as a foghorn; glints of stainless steel flashed off a slate-blue flank that arched above the water. A finback, large and fast, passed us like a bullet train, ending with a relatively tiny dorsal fin. There in the fluke print, a pinkish plume: krill.

Sperm whale fecal plume in the South Pacific. (courtesy Tony Wu)

The smell varies too. When feeding on fish, humpbacks and fins have poops that smell relatively mild, like brine with a hint of sulfur. Poops from krill or copepod meals are stinkier, and the plumes of the North Atlantic right whale are the foulest of them all, a stench so thick that if it gets in your clothes, it'll never go away.

And when we looked at the feces, what did we find? There were high levels of ammonia, a nitrogen compound found in animal waste, in all the fecal plumes. This nitrogen could be picked up by phytoplankton. There were also high levels of phosphorus, another essential nutrient. Whales are common in the gulf in summer, which is when it matters most. The pump fertilizes phytoplankton when the

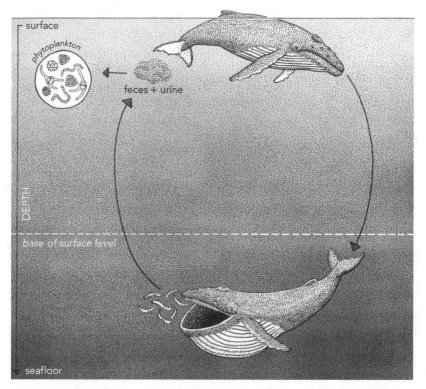

The biological pump is a mechanism for exporting carbon and other nutrients to the deep sea; it has been recognized since the 1970s. In contrast, the whale pump, shown here, is a relatively new concept, in which marine mammals and other air-breathing vertebrates bring nutrients such as nitrogen, phosphorus, and iron to the surface, where they are available to phytoplankton and other primary producers.

thermocline keeps many of the nutrients locked away in the depths, the algae feed the zooplankton and fish, and they in turn feed the whales. It's a positive feedback loop: more whales lead to more fish.

The life cycle is fundamentally the same for any creature. If you follow an animal long enough, it will poop. Eventually, it will die. But even when you follow whales all day long, their feces can be a challenge to find and an even bigger challenge to collect. One afternoon, a fin whale swam toward us off the island of Grand Manan in the Bay of Fundy. Just before the whale dived, I saw a shimmer of white: the scales from a recent fish meal. By the time we got there, it was gone. The shit that got away!

So how to improve our chances? About twenty years ago, my colleagues Roz Rolland and Scott Kraus of the New England Aquarium had a crazy idea. They had been measuring hormones in whale poop to assess stress levels and reproductive status (that is, doing fecal pregnancy tests) for a couple of years, but they were having a hard time finding samples. Scat-detecting dogs were employed to find poop from grizzlies, mountain lions, and rare African wild dogs on land—could these dogs help find whale scats on the water?

Rolland and Kraus rang up Barbara Davenport of PackLeader, a dog-training organization, and asked if she had any detector dogs available for the summer. It turned out she had two. Fargo was a ninety-pound six-year-old Rottweiler. He had started out as a companion dog for an elderly woman, but when she went into a nursing home, he began working as a sniffer dog for Davenport, who was employed by the Washington State Department of Corrections at the time. Fargo specialized in cocaine and narcotics for a while, then he landed what must have been his dream job: wildlife-detector dog. He had found plenty of scats in the mountains and forests of the West, but he had never worked at sea. His colleague, Bob, a mixed breed rescued from the pound, had a great nose and disposition.

How does this scat detection work? On land, dogs follow a scent, reversing course when they lose it, homing in on the target as the odor gets stronger. It's a tougher job than it seems because scents can drop out behind thick vegetation or in rough terrain. It should be easier on open water, since there is nowhere for a scent to hide. Easier for the dog, perhaps, but reading canine body language—lots of nose pointing, posturing, and tail wagging—requires skill, experience, and intuition on the part of the handler. Fargo had a docked tail; his boatmates called it "the happy inch" when he was on a scent.

We soon got a whiff for ourselves. Though it requires the sharp nose of a dog to catch the scent of whale scat from a distance, even humans can pick up the complex bouquet of oil and crushed crustaceans with a hint of sourness and, well, whale within about a hundred yards. Human researchers can collect one or two samples a day by watching whales at the surface, but a dog like Fargo can sniff out one every hour or so, a rate that Rolland referred to as "poops per unit effort." Fargo's reward? A tennis ball and a tug-of-war with Rolland.

Fargo's nose was well proven—one of his most spectacular feats was detecting a right whale's scat from a nautical mile away—but he had one minor flaw: he suffered from mal de mer. Before they headed out, Rolland, a veterinarian, administered twenty-five milligrams of Benadryl to prevent nausea. Bob, Fargo's colleague, had a bigger issue: he was afraid of whales and curled up on the bottom of the boat whenever one surfaced nearby. After a season at sea, he retired to a farmhouse in Vermont.

When the dogs were unavailable, we found that the best approach was to look for a courtship group. Right whales spend much of their time in the summer feeding, either skimming the surface with their baleen, filters the size of a grand piano, or more typically diving deep below, then coming to the surface with mud on their heads. They are mostly looking for *Calanus finmarchicus*, the North Atlantic's most common copepod, a crustacean so small that ten thousand could fit in a soupspoon. But every once in a while, a large group of whales gathers in a courtship, or surface-active, group. You can see them from miles away: blows like tiny silver fireworks on the horizon, flukes and pectoral fins rolling at the surface. In this courtship group, several males, from two to more than forty, jostle to get to a female. For a while—and these groups can last for many hours—the ocean comes alive in a tempest of whales, and you might think back to when such encounters must have occurred throughout much of the North Atlantic. The focal female is often seen floating on her back with her genital slit just above the surface, apparently an attempt to avoid copulation or perhaps to weed out the weakest of the males, who are all holding their breath to get a chance at the female. Occasionally, you see what looks like a sea snake—the right whale penis is one of the biggest in the animal kingdom, nine feet long, and the whale can move it around like a finger.

But we weren't interested in the sex. Often, the female will defecate while she's swimming on her back, erupting in what can only be described as a shit volcano. Whales were scudding in across the horizon. Did the smell attract them too? We eased in with a plankton net and carefully transferred the poop to our collecting vials. We later found high levels of nitrogen and phosphorus in the plume, and when we spiked seawater with right whale feces, the growth rate of the phytoplankton doubled. So there is a connection between sex and productivity—the way right whales do it, at least.

MUCH OF THE research on animal subsidies has been done on nitrogen and phosphorus. Why focus on these two elements? Their presence is limited on the planet. (Nitrogen is quite common in the atmosphere, but it's inaccessible to most plants and animals in that inert, gaseous form.) Without nitrogen and phosphorus, the scaffolding of living systems would fall apart, and the engines of life, mitochondria, found in just about every animal and plant cell, would run out of gas.

We can get oxygen, another essential element, along with hydrogen when we breathe or when we drink water, H_2O. Eating provides fuel for two other essentials of life: energy and raw materials. As heterotrophs, just like almost all animals, we need to consume food, often in the form of simple sugars such as glucose (made of carbon, hydrogen, and oxygen). These sugars hold the electrons that power most of our metabolic pathways.

We also need food to get the raw materials, such as nitrogen and phosphorus, we use for the building blocks of proteins (amino acids), DNA (nucleotides), and cell structures. Let's look at the genetic code. Double-stranded DNA is a long polymer made up of nucleic acids. You can see the recurring elements of life in each of these molecules: a five-*carbon* sugar attached to a phosphate (*phosphorus*-containing) group and a nitrogenous (*nitrogen*-containing) base. Nitrogenous bases come in four different forms: adenine, guanine, cytosine, and thymine. This variation is essential to forming the genetic code and providing the blueprints for proteins. Mutations occur when these nitrogenous bases change.

Your DNA is the same in every cell in your body. It's the proteins—molecular machines with moving parts—that carry out different functions, which vary among neurons and muscle and fat cells. One out of every six atoms in a protein molecule is nitrogen—number seven on the periodic table. All proteins are encoded in the DNA, but they're synthesized in different cells depending on the need. Some proteins are for defense, such as antibodies, which fight disease. Some are for storage, such as hemoglobin, which stores oxygen. And some are for communication—for instance, insulin to regulate blood glucose. Without nitrogen, you wouldn't be able to make a muscle. You wouldn't have enzymes for digestion or for forming blood clots. You wouldn't have hormones for growth and reproduction.

Nitrogen is also an essential component of chlorophyll, the engine of photosynthesis, and the reason most healthy foliage is green. (Plants and phytoplankton are autotrophs, which means they make their own food.) If you're a gardener, you might recognize a telltale sign of nitrogen depletion: yellow leaves. Without nitrogen, plants start to wilt and look sick. In the oceans, when nitrogen runs out, photosynthesis also comes to a halt, and the seas become unusually clear—a nutrient-poor ocean is a transparent ocean and a welcome destination for many vacationers.

HENNIG BRANDT WAS convinced that human urine could be turned into gold. They had similar colors, and like gold, the human body was considered a work of perfection in seventeenth-century Europe. So it was only natural that urine and gold would be connected. In 1669, Brandt, a practicing alchemist, collected fifty buckets of urine from his neighbors in Hamburg—then as now, beer drinkers were a reliable source—and left it to ferment. After heating and distilling the residue, he found a white solid that combusted when exposed to air. He called it phosphorus, "light bearer" in Greek.

If nitrogen is the star of the book, phosphorus, number fifteen on the periodic table, deserves a nod for best element in a support- ing role. It turned out that phosphorus was far more valuable than gold for human life and for the history of life on Earth. Our planet is about 4.5 billion years old. Living organisms showed up, mostly in the single-cell form, about 3.8 billion years ago. It took another billion years or so for phytoplankton to start cranking out oxygen, but there wasn't enough phosphorus to go around until about 750 million years ago, when an increase in the concentration of this vital element bumped up the number of complex life-forms in the oceans.

Changes in the phosphorus cycle occurred during a time of great climactic upheaval, a shift in ocean chemistry, and the emergence of new complex organisms. The first animals, metazoans, were probably relatively simple in the beginning—cells surrounding something like a canal that could trap food and release digestive enzymes that would otherwise drift away. They begin to appear in the fossil record about 600 million years ago, followed by more complex life-forms in the sea and eventually on land. Animals are newcomers in the long his- tory of the planet. They depend on phosphorus and help distribute it.

This essential element provides molecular and organismic scaf- folding, forming the backbone of our DNA. Single-stranded RNA, the link between DNA and proteins, is phosphorus-rich and required for rapid growth. Phosphorus also hardens the skeleton; an average human body has about a pound or two of phosphorus, necessary for strong teeth and bones.

Phosphorus is critical for energy flow; cellular respiration is at the core of all metabolic processes. The food that we eat is transformed into energy in the cell through the Krebs cycle—the movement of electrons from one molecule to another until the energy is captured in a manageable way. This cycle, the essence of life, occurs in the mitochondria, the organelles and energy powerhouses of eukaryotes, complex organisms that include plants, animals, fungi, and protists. If life in the cell is a "molecular storm," as one biologist described it, this pathway helps establish a sense of order. The power from transferring electrons is harnessed to add a phosphate group to ADP, adenosine diphosphate, turning it into a triphosphate, ATP. When the molecule has three phosphates, it has stored energy. The release of a single phosphorus atom between ATP and ADP provides a jolt of energy, fueling all of life. This is carefully calibrated and incredibly common. "On any given day you turn over your body weight equivalent in ATP," the chemists Susanna Törnroth-Horsefield and Richard Neutze wrote.

"Is it ecologically important or a fart in a hurricane?" Dick Barber, a biological oceanographer, once asked me over lunch at the Duke Marine Lab on the coast of North Carolina. We had been discussing the mechanism behind the whale pump. (Biologists don't mind talking about the process at the end of the digestive system while involved in the process at the beginning.) A classic bottom-upper, Barber could see how a fecal plume might fertilize the surface of the sea, but at what scale?

That plants, bacteria, fungi, animals, and other organisms play a role in the biogeochemical cycles is not in question. But the focus has often been on the small things. Hundreds, maybe thousands, of studies have examined phytoplankton's role in the foundation of marine life, in climate regulation, even in cloud formation. These bottom-up studies might also look at upwelling, light, currents, or temperature. Top-down studies of animals—organisms that eat other things to survive—frequently focus on browsing, grazing, and predation. But the biogeochemical role of these animals has often been ignored. The effects of rotting carcasses, deposited feces, and other unglamorous forces in dispersing, concentrating, recycling, and moving nutrients around local ecosystems, not to mention their impact on the global climate cycle, have been largely overlooked.

This is starting to change, with new concepts being tested and new terms coined. There's the whale pump, as we just explored; the hippopotamus conveyor belt, which we'll discuss soon; and *zoogeochemistry*, which places animals at the center of biogeochemical processes. The dynamics that shape our physical world—atmospheric chemistry, geothermal forces, oceanography, plate tectonics, and erosion through wind and rain—have been explored for decades, even centuries. The evolutionary consequences of competition and predation have been contemplated at least since Darwin published *The Origin of Species.*

"How did we not know this?" I've been asked over the years. How did we not see that animals—from fish to seabirds, whales, and bears—could shape ecosystems, from the deep sea to the mountaintops? I have a few ideas.

First, we live in a world where wild animals have largely disappeared from view, so it's easy to overlook their historic roles. On land, animals make up only about 5 percent of the total biomass; the rest is mostly plants. Walk out your door, and you'll surely see more trees and grass than animals. But in the ocean, animals, many of them invertebrates, outweigh plants five to one. Yes, it's easy to think that whales aren't important when we've removed about two-thirds of them from the ocean. Before humans started hunting them, there were more than four million whales in the oceans. Now—after centuries of harvest and a few decades of protection—there are about one and a half million. Many species persist in relict populations, such as North Atlantic and North Pacific right whales. Fully 99 percent of blue whales in the Southern Ocean were killed in the twentieth century. Humanity has even reduced the size of whales by almost a third by killing the largest ones first. The biggest animals that ever existed on the planet have been whittled down from an average size of ninety feet to seventy-two. Modern-day blue whales are shorter than their ancestors by the length of a giraffe.

Second, there is an implicit bias in much of biology supporting bottom-up processes, an acknowledgment that plants, phytoplankton, and microbes are the foundation of the world. For sure, we, along with the vast majority of animals, couldn't live a day without plants. But they don't tell the whole story. Many plants would be lost without animals. Most flowering plants rely on animals for pollination; others use the nutrients from their feces, urine, and carcasses. Some thrive when browsers or grazers reduce otherwise unvanquishable competitors; others rely on birds and mammals to move seeds, sometimes to islands as far-flung as Surtsey. The relationships are numerous and complex. As ecologist Jim Estes, whom we'll meet later, noted: In a community of just a hundred species, there are billions of billions (2.5×10^{157}, to be more precise) of potential direct and indirect interactions. At best, we'll uncover a few of them.

Third, population biologists, scientists who study individual species and population dynamics, don't tend to collaborate with ecosystem ecologists, who study biological and geological processes like nutrient cycles and energy flow. This is unfortunate. Biologist Paul Ehrlich recognized an urgent need for integration in the 1980s, and Clive Jones and colleagues coined the term *ecosystem engineers* in 1994, later publishing a book called *Linking Species and Ecosystems* that explored the ecological roles of beavers, sea otters, and even organic detritus like marine snow. This is all to say that only within the past three decades or so have the ideas of ecosystem engineers and nutrient subsidies made their way into the ecological literature. This is a welcome development, but the chasm persists.

In part, this is because biological disciplines are rarely inter-disciplinary. Ecosystem ecologists and population biologists go to different meetings, submit to different journals, address different questions. A fish biologist might claim that oceanographers over-look anything that the eye can see in their pursuit of the microscopic processes that dominate the ocean. An oceanographer could claim that fisheries biologists don't really study science but vainly pursue maximum sustainable yield, chasing after bigger catches and ignor-ing basic biological processes. I've heard such claims over beers (at best) and Zoom (alas). Evolutionary biologist David Sloan Wilson observed, "The Ivory Tower would be more aptly named the Ivory Archipelago," with "hundreds of isolated subjects, each divided into smaller subjects," each with its own history and assumptions.

On occasion, a few have breached these barriers. Evolution can unite our understanding of the world, from microbiology to brain science. Ecology can link biological and evolutionary processes with physical and geological ones. George Evelyn Hutchinson, a towering figure in ecology, was one of the first to collect and syn-thesize evidence for the importance of bird guano. In *The Biogeo-chemistry of Vertebrate Excretion*, a 570-page monograph published in the *Bulletin of the American Museum of Natural History* in 1950, he built the case for preserving and encouraging large bird colo-nies by showing that seabirds, rather than competing with people for food, could actually increase fish populations by enhancing nu-trients close to their nesting grounds. An earlier paper, written in 1923 by English zoologist Charles Elton and botanist V. S. Sum-merhayes, investigated the role of guillemots and gulls on the bird cliffs of northern Norway. Seabirds appear to be the ambassadors of resource subsidies.

In the 1970s, a few ecologists began examining the shifts in ecosystems when animals appeared or disappeared, but it wasn't really until the past decade or so, with the emergence of zoogeochemistry—which explores how animals influence the flow of key elements, such as carbon, nitrogen, and phosphorus, through living systems and the physical environment—that we've seen a shift in the understanding of the role played by animals in altering landscapes and seascapes. As birds and whales do at sea, bison build grasslands by choreographing the green wave, spiders boost primary productivity in meadows by spinning a landscape of fear, and sea otters build kelp forests by keeping sea urchins at bay. We'll revisit these species soon.

WE HAVE BEEN talking a lot about nitrogen in this book. It is often the limiting element in new landmasses and in coastal systems of the Northern Hemisphere. But things play out differently in other parts of the world. The Southern Ocean can be nitrogen- and phosphorus-rich, but it's limited by an element that is less abundant in plants and animals.

"Iron is absolutely crucial," the marine scientist Victor Smetacek told me over Skype from his home near Bremen, Germany. Along with nitrogen and phosphorus (the macronutrients), iron is one of the essential elements of life and quite common in many places, forming much of Earth's core. Although it is needed in smaller amounts than the other two—it's considered a micronutrient—it is essential for synthesizing chlorophyll, so phytoplankton and plants can't photosynthesize and grow without it. Iron also plays a key role in delivering oxygen via hemoglobin in many animals.

In 1990, oceanographer John Martin proposed the iron hypothesis, a controversial capstone to a prestigious career in ocean science. He noted that much of the vast Southern Ocean had plenty of nitrogen and phosphorus but still had low productivity because it was iron-limited. Iron enrichment, from atmospheric iron dust, had enhanced the growth of phytoplankton in an earlier period. He theorized that if we wanted more productivity, seeding the Southern Ocean with iron could boost the growth of phytoplankton, with global consequences for carbon drawdown. "Give me a half tanker of iron," he once said at Woods Hole, "and I will give you an ice age." Little did he know that, thirty years later, with the widespread melting of glaciers around the world, many people might see that as a good thing.

Smetacek recalled mocking Martin at a later meeting, calling him Popeye for his dedication to iron. Others called him Johnny Ironseed. "At the time he proposed his ideas," Smetacek said, "I didn't believe him." But as Smetacek looked at iron in the ocean, much of it locked away in biomass, he started to change his mind. "And then Martin passed away, which made me feel really bad, because I wanted to tell him that he was right and I was wrong," Smetacek mused. "That's why I understand so many of my colleagues who are still skeptical about the role of iron."

In the 1930s, when whales were still relatively common in the Southern Hemisphere, British scientists remarked on the astonishing abundance of life in the Southern Ocean. "They were seeing huge amounts of diatoms all over the place being eaten by krill," Smetacek said. Diatoms, unicellular algae, are large and numerous, among the most common organisms on Earth. By the time Smetacek got to the Southern Ocean in the 1990s, whales had mostly disappeared; the biggest species, Antarctic blue whales, were down by 99 percent. Many scientists were focused on the microbial loop, the ecology of microorganisms as the predominant process in offshore systems. The small whales that were left, such as Antarctic minkes, were hardly ecosystem players at all.

What had changed? Smetacek, who had so cavalierly dismissed the iron hypothesis early in his career, started to come around to the idea that iron was more important than he'd realized. The more he read and the more time he spent on the water, the more he understood that the absence of this micronutrient in the Southern Hemisphere could lower the growth rates and biomass of diatoms and other phytoplankton. In northern oceans, iron is rarely limiting, since rivers and desert dust blowing north over Africa are often rich in the element. He became convinced that iron could boost the growth of phytoplankton.

As it turns out—as I'm sure you've guessed by now—whale fecal plumes are a natural source of iron. In addition to being rich in nitrogen and phosphorus, the concentration of iron in whale poop is more than ten million times greater than in the surrounding seawater in the Southern Ocean. Whale poop once provided tens of thousands of tons of this micronutrient to the surface waters in the Southern Hemisphere; shallow-feeding whales recycled it, and deep-diving sperm whales brought it up from the depths. As the whales disappeared, so, too, did an important source of iron. Productivity declined and ecosystems shifted. According to Smetacek, phytoplankton, krill, and whales all suffered when the positive feedback loop was broken.

Smetacek has become one of the main proponents of artificial iron fertilization; adding iron sulfate to the ocean surface in areas such as the Southern Ocean, where iron rather than nitrogen or phosphorus is the limiting nutrient, will boost productivity, and some of that carbon will sink through the biological pump, helping to cool the planet. Iron fertilization could even help increase the number of krill and whales, revitalizing the whale pump in the Southern Ocean.

"We did the last experiment in 2009, when we were putting tons of iron out in the ocean," Smetacek said. He pulled out a green plastic tub containing about ten pounds of iron sulfate, used for fertilizer on land. "It's exactly the stuff we were using for iron fertilization at sea. This is almost enough iron to feed a blue whale, right?" His dream was to use this fertilizer to re-spark the abundance of the Southern Ocean; if you added iron and restored whales and the whale pump, the diatoms, krill, and large vertebrates would all benefit. And it would help in the fight against climate change too, because some of the carbon would eventually sink into the deep sea in the form of dead plankton, dead fish, and dead whales.

"You know, I told that story to Jim McCarthy," Smetacek said. He had gotten a lift with McCarthy to a meeting I was hosting. Though perhaps best known for his work with the Intergovernmental Panel on Climate Change, McCarthy—who, sadly, died in 2019—was a biological oceanographer who had studied marine nitrogen in the Northern Hemisphere since the 1960s. Smetacek had been working in the iron-limited Southern Ocean for almost as long.

They'd chatted about oceans, climate change, and personal health during the long car ride. As a universal donor—a person with type O blood—McCarthy had donated blood three or four times a year for much of his life. When he was in his sixties, he was diagnosed with hemochromatosis, a disorder in which too much iron is deposited in the tissues; it causes damage to the heart, liver, and other organs. McCarthy, who had studied nitrogen-poor (and iron-rich) marine areas for much of his life, had too much iron in his body. One treatment for the disorder is drawing blood several times a year, and Smetacek pointed out that McCarthy's donating blood so often might have saved his life.

"Two or three years ago, I was diagnosed with iron deficiency," Smetacek told me. He had recently become a vegetarian. "I have to take the same stuff that we release at sea, iron sulfate." Like the ocean he studied, Smetacek was iron-limited.

As SUMMER COMES to an end, great whales start to move, leaving their foraging areas in the high latitudes of Antarctica, Iceland, and the Gulf of Maine and heading to warm-water calving grounds closer to the equator. In the summer when they're feeding, humpbacks can spread out as if grazing at an enormous seasonal buffet. In Alaska, they might start with some krill in Sitka Sound, move on to the capelin of Chatham Strait, then settle down in Glacier Bay to bubble-feed on herring. Whales are capital breeders—animals that use stored energy, or capital, to "finance" reproduction. In the whales' case, that capital is blubber, so they bulk up in the productive summer months, then travel to warmer low-latitude waters to give birth, nurse, and mate. They aggregate on these winter grounds—you can eat alone, but you need a mate to breed.

My colleagues and I have worked on the vertical movement of nutrients—the whale pump—but whales also move great distances across Earth. Gray whales migrate more than six thousand miles, from Russia to their breeding grounds in Mexico. Southern Hemisphere humpbacks travel similar distances between the Antarctic and Samoa.

Why do whales migrate such vast distances to breed? It's still not entirely clear. Many of my colleagues give the credit to killer whales, the great whales' only lethal marine predator. Calves are vulnerable to attack by orcas, and low-latitude sheltered areas may offer refuge and an advantage in active physical defense, especially for slow-swimming species. (Shallow water makes it more difficult for killer whales to attack from below, as they do in deeper seas. It also muffles sounds, since vocal signals don't travel as far. Moms and calves can stay in touch without ringing the dinner bell for off-shore predators.) Another reason whales leave higher-latitude, colder waters in the winter and travel to areas where they are thermoneutral might be to conserve energy. This hypothesis also applies to neonates—calf development in warm water can lead to larger adult size and increased reproductive success.

Bob Pitman of Oregon State thinks it's like going to a spa. In warmer waters, whales can slough off their skin and rid themselves of fouling organisms such as microscopic diatoms that form thick yellow-green layers in the nutrient-rich higher latitudes. They're exfoliating all the time, Pitman says, releasing the microbes growing on their bodies into the warm ocean.

The crystal-clear waters of the tropics are low in nutrients—great for visibility but not for productivity. The whales don't mind, since they don't tend to feed in the winter, but they can deliver nutrient and food subsidies. When they're fasting, they're living off the blubber and muscle stored during the summer. The nitrogen and phosphorus from the breakdown of these stored materials have to go somewhere. It's not just in carcasses or poop that these nutrients move around the world; it's also in pee. In fact, most whales probably don't defecate on the breeding grounds—if they're not feeding, they aren't pooping.

They might not be eating, but they are metabolizing, burning energy and releasing nitrogenous wastes in the form of urea. It's a struggle to collect whale poop in the field, so forget about pee. For the most part, we have to rely on metabolic models and our understanding of other capital breeders, such as elephant seals, to figure out when and how much whales urinate.

ONE WEEK IN mid-February, I visited whale biologist Chris Gabriele and her colleagues on the Big Island of Hawaii to see a few humpbacks. We sat along its sandy edge on a tuft they call Old Ruins. There was *vog*, volcanic fog, to the south, drifting in from Kilauea. For much of their time in Hawaii, whales can be found in shallow, sandy waters not far from shore. The whales here have one of three goals: Get pregnant. Get someone pregnant. Or (for the moms with new calves) be left alone.

One mom swam north with her newborn calf at her side. We saw her fluke hit the ocean surface, then heard the slap moments later. We counted five pods that day, as well as a whale-watching boat, a dive boat, and several other vessels. I was embedded as a notetaker. Although we saw only a handful of whales and a bunch of spinner dolphins, I struggled to keep up.

It was much easier when they started, Gabriele told me. Humpbacks were still endangered during their first field seasons, in the 1980s. Whale numbers started increasing after the Russians stopped hunting them in the North Pacific, then peaked at around twenty-one thousand in 2013. There were so many whales in the waters that they occasionally saw as many as twenty-seven pods at a time. Gabriele mentioned that after years of conservation success, humpback whales were now declining in the North Pacific.

"We stopped killing them," Adam Frankel, one of Gabriele's colleagues at the Hawaiian Marine Mammal Consortium, said from behind the binoculars, "and now we're killing their climate."

"And now we're killing their food," Gabriele added.

"No rest for the incredibly weary."

It seemed that twentieth-century whale conservation was almost too easy: race a Zodiac between a whaling ship and a sperm whale, make a splash in the media, and keep fighting whaling one ship at a time. Today, the battles are so much larger; we're watching humpbacks and other whales suffer from a lack of fish and zooplankton as warming waters destroy their habitats. We are all whalers, as marine scientist Michael Moore has written, even if we do it inadvertently, through ship strikes, fishing entanglements, and polluted oceans.

The following day, Gabriele's small research boat was left on land due to high winds, so I hopped in a large Zodiac with a bunch of civilian whale watchers. We set out on shockingly blue waters offset by the dry yellow and green hills. I had spent years listening to the relatively simple "up" calls of right whales on their feeding grounds in the Bay of Fundy, but this was my first time in a humpback breeding area. One of the crew dropped a hydrophone in the water. Even though the weather was rough and the whales skittish, the sound was so clear that at first I thought it was a recording. The ocean was resonant with a deep, shore-to-shore sound. It was like being transported to a concert hall, hearing songs with the showiness of an acoustic peacock tail, as one biologist described it. An unexpected sense of awe.

After a day of watching whales, I joined Gabriele and her husband, Paul Berry, at a nearby restaurant. There was a long wait, so we walked to the water and sat on a broken-down seawall. The sun slipped below the horizon.

And then it happened: a brief green ray above the sun.

"Did you see that?" I asked. It was an emerald flash, a momentary tower of neon just above the sun as it sank below the horizon.

"It's more like a green moment than a come-to-Jesus flash," Berry said dryly.

Many people, including myself, have worked on the ocean for years and never seen the flash. If there had been fewer diners that evening, we would have missed it.

***

As I WAS out on the water, my colleagues at the University of Hawaii were placing temporary suction-cup tags with cameras on the backs of humpback calves off the coast of Maui. Later, we would see a green curtain of pee—a calf's-eye view of mom urinating in the baby-blue sea. The mom didn't feed in Hawaii, but she certainly released nitrogen and phosphorus there. A few mackerel scad picked off Alaska-grown whale skin and hovered around the nursing whale in search of extra milk, nutrient transfer at work.

What does the movement of these nutrients mean for Hawaii and other tropical and subtropical areas? Humpbacks transport about a million pounds of nitrogen to Hawaii each winter. That's more than twice the amount of nitrogen released from physical mixing—wind and upwelling, the traditional focal points of many oceanographers—into the Hawaiian Islands Humpback Whale National Marine Sanctuary. This imported nitrogen is a significant source of nutrients for phytoplankton, and they can absorb thousands of tons of carbon dioxide from the atmosphere. And it happens all over the world. Whales transport more than thirteen million pounds of nitrogen to their breeding grounds in the Northern and Southern Hemispheres each year. To our knowledge, this movement represents one of the largest long-distance animal subsidies on the planet.

This great whale conveyor belt, as we called it, delivers much more than pee. Pregnant females bulk up for the long, mostly foodless winter, then release nutrients and food in the form of offspring, placentas, and milk. Calves poop on the breeding grounds after eating about a hundred gallons of fatty milk a day. Whale placentas can weigh up to fifty pounds. As it is for many other mammals, infant mortality is high just after birth. A newborn calf is enormous—twelve feet long and about three thousand pounds—so a calf carcass releases a large amount of food in the form of biomass and nutrients into the low-latitude ecosystems. A white shark can survive for six weeks after feeding on sixty pounds of whale blubber.

Along the coasts of Brazil and Australia, where shark nurseries overlap with humpback breeding areas, tiger sharks feed on calf carcasses and occasionally bite young whales.

Adult whales have higher rates of survival, but everyone has to die. Given the competition on the breeding grounds and the complications of birth, I wouldn't be surprised if whale death turned out to be more common in the winter, providing an enormous burst of nutrients and food. The great whale conveyor belt delivers unprocessed energy- and nutrient-rich meals in the form of blubber, muscle, bone, organs, and placentas straight to the sharks, groupers, and other denizens of the reefs around the islands. Altogether, humpbacks provide more than seven thousand tons—the weight of twenty-nine million Big Macs—of biomass to Hawaii each year through their placentas, carcasses, and shedding skin.

The females who get pregnant leave Hawaii first. They've achieved their goal on the breeding grounds, and they head straight for the all-you-can-eat buffet in Alaska in late February. The juveniles, small, hungry, with no reproductive need to stick around, leave soon after. The females with calves depart later, generally in March, so they can give their calves time to grow before the long swim north. Some males might stick around the breeding grounds for a little longer on the off chance that they'll find a receptive female before they head north. All these animals have been through a lot—a long period of fasting, then fighting for breeding opportunities or the stresses of giving birth and nursing—and a few die of exhaustion or old age on the way back to the feeding grounds. And when they do, well, it's a different kettle of fish.

All great stories start off with a dead body. What about one the size of a school bus?

IN 1987, CRAIG Smith was a postdoc in charge of running transects through Catalina Basin off California with *Alvin*, the first deep-sea manned submersible. One afternoon, his grad students came across a long stretch of bleached bones surrounded by pink worms and big clams.

"They used the underwater telephone to call up to the ship," Smith, in a pale blue aloha shirt and a San Juan Island baseball cap, told me over dinner in Honolulu. "They said they had found something interesting and that they were going to spend a little more time on the bottom." They came to the surface with a surprise—a big vertebra in the basket of *Alvin*—and some images of the seafloor. The video showed large white clams that had previously been found only around hydrothermal vents. In these ecosystems, organisms depended on sulfur and the hot waters emitted from the vents to survive. What were they doing around the skeleton of what looked like a whale?

Smith, now a University of Hawaii professor emeritus, and colleagues later described this new deep-sea habitat as a *whale fall*: a carcass that drops to the ocean floor, typically more than a mile below the surface. "One of the things that's important to consider about whale falls," he told me, "is that they're sinking into a very food-poor environment." Unlike the high-latitude upwellings where many whales feed, the deep sea has very little food due to the absence of light and photosynthesis. Most of the organic material in the deep sea rains down from the ocean surface in the form of minute particles, decomposing cells, microbes, and brightly colored aggregations known as marine snow. The abyssal seafloor is a vast nutrient-poor desert.

For hundreds of species—sleeper sharks, deep-sea octopuses, zombie worms, tiny amphipods, enormous crabs, sea anemones, and species we haven't even discovered yet—a dead whale is both prime real estate and a buffet that might last for years. When the carcass of a big whale hits the seafloor, it's like a blizzard of marine snow, about a thousand years of biomass in one big fall. "It's a huge, immediate pulse of energy-rich material: food, lipids, proteins that scavengers can use," Smith said. The soft tissue provides food, and whale skeletons—unlike, say, those of whale sharks and other large fish—are robust and high in minerals. Whale bones are up to 70 percent fat and are set in a mineral matrix so tough that only microbes can enter through the porous spaces. "The skeleton forms an organic-rich reef that's very slowly releasing this sulfide- and energy-rich compound," Smith said. "These compounds can be used by free-living microbes and microbes that live within the tissues of clams, mussels, and tube worms—endosymbionts." At times, the bones look like they're covered in slow-moving curtains, a thin veil of worms.

"We collected them in 1996, and nobody knew what they were," Smith told me. "We were still calling them snot worms." He showed these novelties to several polychaete experts, invertebrate biologists who specialized in worms, but they weren't sure if they were a unique branch of the marine annelids or a distant relation. Polychaetes are a diverse group of mostly marine species that include showy shallow-water filter feeders, such as the descriptively named Christmas tree worms and feather duster worms, and deep-sea hydrothermal tubeworms that rely on symbiotic bacteria for sustenance. Even Adrian Glover, the polychaete guy working in Smith's lab, couldn't figure out how these strange bone-eating worms fit in with the rest of the deep-sea fauna.

But then researchers from Monterey Bay Aquarium Research Institute collected a big specimen from a dead gray whale off Monterey in February 2002. They could see the anatomy of the snot worms much better. After comparing the body shape and the DNA to other marine worms, they identified it as a polychaete, one that seemed entirely dependent on dead whales. They gave it the genus name *Osedax*, Latin for "bone-eater."

Despite its scientific name, *Osedax* is an outlier among the animals in this book's terrain. It doesn't eat and it doesn't poop (but everything has to die). *Osedax* has no mouth. It has no gut. It has no anus. These gutless worms have a colony of bacteria in their roots that can decompose lipids and proteins in the whale bones. The root structure releases acid that penetrates the bone, anchors the worm, and allows it to take in nutrients. Along the surface, plumes act as gills, a red feathery boa flowing above the bones. Skeletons with *Osedax* are species-rich, the habitats complex and changing over time.

Osedax lives on whale falls in ways that are analogous to how animals survive on hydrothermal vents and cold seeps. Many animals in these deep-sea ecosystems rely on microbes to survive. If the sun craps out, hydrothermal-vent communities, which depend on chemosynthesis from sulfides released from vents rather than on photosynthesis, have a good long-term strategy. On the whale falls, *Osedax* and other chemosynthetic animals absorb the fats and sulfides in the bones, and their internal bacteria synthesize nutrients and release them back to the animals.

Since the first *Osedax* was described, back in 2004, more than thirty different species of these worms have been found on whale falls. "Some of these are whale-fall specialists," Smith said. "They're just too big to live on the bones of other animals." *Osedax* arrives early and often to these carcasses, densely colonizing the whale bones. After months of pulling the sulfides out of the skeletons, the bone-eaters spawn and release larvae into the deep, dark water column. They rain down, these derelicts of larval dispersal. The lucky ones find a new whale carcass—the next generation of whale-fall connoisseurs. Most miss their mark and never find a suitable habitat, a lonely early death in the ever-dark.

Those marks have gotten even harder to find in the past hundred years or so. When whales were rendered on sailing ships, their carcasses were often dropped overboard after the blubber was removed. Melville described the removal of the outer layer of fat as being like peeling an orange, "the whale rolling over and over in the water" as "the blubber in one strip uniformly peels off." The early peak of whaling must have been a boom time for many deep-sea creatures—dead bodies were all over the bottom of the ocean—but as the largest whales were lost to harpoons, these habitats began to get smaller. Later, when whales were hunted by factory ships, their entire carcasses were cut up on the flensing deck. The whale-island habitats for these deep-sea creatures went up in smoke or were rendered into dog food or fertilizer.

At least one hundred species that appear to be whale-fall specialists, unique to the enormous carcasses and bones, have been discovered in the deep sea. Many of them are found in the sulfophilic stage, after most of the soft tissue has been eaten away and the nutrients are primarily coming from the bones. More than forty thousand individual animals were found on a single whale seventy years after its death—a small whale-fall city of deep-sea isopods, polychaete worms, and tiny translucent clams representing more than two hundred species, some specific to whales, others deep-sea generalists. But as their real estate disappeared after decades of relentless commercial whaling, whale-fall communities started to vanish too—first individuals, then populations, then species. Some of the earliest extinctions in the ocean were probably whale-fall specialists, left without the enormous habitats they had relied on for millions of years. Homeless, they blinked out. Gone forever.

Along with declines in deep-sea diversity, with fewer whales, there were disruptions in the movement of nutrients from the seafloor to the surface and from high to low latitudes. There has also been a climate impact of whaling. When whale carcasses sink to the deep sea, their bodies can sequester carbon for centuries or even longer. Commercial whaling broke that cycle by rendering bones and burning the oil in blubber, releasing it into the atmosphere. Before whaling, whale falls sequestered more than six million pounds of carbon each year. That has been reduced by about two-thirds. Nowadays, many whales die tragically—starved by nutrient-empty oceans, entangled in fishing lines, or struck by ships—and in those cases, their carcasses often wash up on the shore. Too few die of natural causes and sink to the deep.

As GLOBAL TRAVELERS, whales link the oceans—deep to shallow, the poles to the tropics. But they also connect the sea to land. When whales strand, they can be a nutritional boon to terrestrial animals. Bald eagles and ravens peck at the dead animal's skin. Wolves feed on the organs. Crabs burrow in the crevasses opened by larger scavengers.

Kristin Laidre, a polar researcher at the University of Washington, told me about a photo taken by some tourists in the Russian Arctic a few years ago. "They stumbled onto a bowhead whale that had died and washed up on Wrangell Island," she said. There were several hundred polar bears aggregated around the whale and extending up to the mountain, like sheep dotting the English countryside. "It was just this completely crazy landscape of polar bears feeding on that whale through the open-water season." This is a time when bears often go hungry, since they can't go out on the ice and hunt seals, their preferred prey. "And so I looked at those photos and I thought, *What does a dead bowhead whale mean to polar bears and how can that translate to getting some bears through a long ice-free season when there's really not much to eat?*" When Laidre dug into the literature, she realized that whales were reliable resources for bears; they had massive amounts of nutrients and lipids that allowed polar bear populations to persist through the long summers and perhaps even during periods of deglaciation. In the past, stranded whales might have been essential to polar bear survival. Perhaps in a warmer future, they will be again.

Further south, California and Andean condors rode the thermals of the eastern Pacific, relying mostly on good eyesight to find their next meal. With a wide distribution, from California to Florida, and a nine-foot wingspan, the California condor was an apex scavenger. In the past, it preyed on the large wild herbivores of North America, but as mastodons, mammoths, and, eventually, even bison disappeared, condors became dependent on whales and other large marine mammals along the Pacific Coast.

Marine mammals declined after the onset of industrial hunting in the nineteenth century, and the carrion birds of South America were left to feed on guanacos—small wild camelids related to llamas—and sheep, horses, and cows if they could find them. In the

North Pacific, the switch from surf to turf nearly caused the extinction of the California condor. With many marine mammals hunted out during the nineteenth century, California condors switched to cattle and deer carcasses, which were often riddled with lead bullets. By the time ornithologist Roger Tory Peterson caught up with one in 1953, California condors were rare, and seeing one was the highlight of a cross-country trip: "It was huge, black, pale-headed. . . . It made a couple of flaps, as if it had all the time in the world, caught a new thermal, and soared away to the southeast until it became a tiny speck and disappeared."

They were all disappearing. Many California condors died of lead poisoning, and in 1982, it seemed only a matter of time before the entire species, down to just twenty-two individuals at that point, would vanish. Every condor was removed from the wild in the 1990s. New generations were raised in captivity, and there are now more than three hundred of these huge free-flying scavengers soaring from Baja California to the Pacific Northwest. The condor is sacred to the Yurok tribe in Northern California, and the first wild condors in the area in over a century were released from tribal lands in 2022. Even now, about 20 percent of free-flying condors have lead levels so high that they must undergo chelation therapy to remove the toxins, which often requires weeks in captivity. (Efforts to ban lead-based ammunition have faced strong opposition from the gun lobby and have so far failed in California and on the federal level.) As marine mammals recover, the carcasses of whales and seals will be essential to the next phase of the condor's comeback.

FOR CENTURIES, CETACEANS that stranded on British shores were known as royal fish and considered the property of the Crown; the oil was used as lamp fuel, the bones for tools. In Denmark, the law was similar, but the person who discovered the whale was permitted a share (as much as he could carry if he was on foot, a larger amount if he was riding). The king kept the greater part, since he owned the shore. In the Bering Sea, a drift whale was considered a gift from Sedna, the goddess of the sea. Australian Aborigines believed that strandings connected them to ancestral lands and seas. In English, the word *windfall* comes from fruits felled by a gale that were free for the taking. For Icelanders, *hvalerecki*—"whale stranding"—is an unexpected fortune that washes up onshore, a gift of meat, baleen, oil, and bone.

After whale meat fell out of gastronomical favor in Europe, however, strandings came to be thought of as a sign of God's anger, a presage of disaster (though it appears that human impact on whales was the true calamity). Today, beached whales are often considered an inconvenience, perhaps a health hazard. What to do with these stranded whales, living or dead? The living are released back into the ocean if possible; if not, they're euthanized.

A dead whale rotting on the shore is considered a public health risk. What do you do with the carcass of a thirty-ton animal when it washes up on a public beach? In its pamphlet "Obliterating Animal Carcasses with Explosives," the U.S. Forest Service recommends three pounds of explosives for a thousand-pound horse, cautioning the blaster to first remove the horseshoes "to minimize dangerous flying debris" and double the amount of explosives for "total animal obliteration." Someone should have warned the Oregon State Highway Division that this might not scale up before they decided to obliterate the beached carcass of a forty-five-foot sperm whale in 1970. Half a ton of dynamite was placed on the leeward side of the whale in hopes of dispersing the carcass in seagull- and fish-size pieces out to sea. The explosion showered blubber and whale guts over the assembled crowd. One large slab smashed through the windshield of a car.

Nowadays, people chop them up, tow them out to sea, or bury them. The bones can be unearthed later and displayed or stored for scientific studies. (The Smithsonian has the remains of more than ten thousand whales, the largest collection of whale bones on Earth.) When a body washes up and we don't know the cause of death, why not just leave it alone? For condors and polar bears and, most likely, many other mammals, birds, and terrestrial invertebrates, the nutrient subsidies from sea to land by stranded cetaceans are enormous. About one out of every four whale carcasses are left in place in the United States—no doubt far from the madding beach crowd.

In much of the world, a dead whale was an occasional event, a cause for wonder and celebration, but what if there had been a regular supply of carcasses, smaller but far more dependable, every year? How would that have shaped the coastal communities, forests, and ecosystems that lined an entire ocean basin?

3

Eat, Spawn, Die

Picture the classic forests of Alaska and the Pacific Northwest: Sitka spruce, balsam poplar, western hemlock. A snow-covered mountain in the distance. The dense woods running from Alaska to Northern California form the largest temperate rainforest in the world. But how do these forests work? Where do the nutrients that allow them to be so productive and have some of the largest trees on the planet come from? The classic idea is that nutrients are recycled in the leaf litter—mulch, carcasses, insect frass—near the roots. Fungi and bacteria, the major decomposers, break down the leaves and dead trees. Over time, nutrients tend to move downhill, running from the mountains to the streams to the ocean.

Back in the 1980s, Bob Naiman, an ecologist in British Columbia, thought there had to be more to the story. The typical paradigm for river ecology is a one-way street: water, nutrients, and other matter go downslope (or downstream), usually winding up in marine ecosystems. Many trees are close to streams that run down to the ocean, but seawater can't run uphill. Naiman, a river ecologist, had observed one huge upstream pulse coming from the oceans each year: spawning salmon.

The classic salmon of the Pacific Northwest—Chinook, chum, coho, pink, and sockeye—hatch in streams, where they spend the early days of their lives eating the nymphs and larvae of stone flies, mayflies, caddis flies, and other invertebrates. After a year or two, maybe three, they migrate out to the ocean. Most of what makes salmon comes from the ocean; they accumulate over 95 percent of their weight at sea. For sockeye, that might be up to ten pounds; for Chinook, it's up to thirty. Many salmon feed on shrimp and krill—which help give them their distinctive pink flesh—and forage fish like anchovies, herring, and capelin.

To Naiman, watching the coastal animals on Vancouver Island, it was clear that the bears and eagles depended on the summer salmon run for sustenance. But he wondered if the migratory fish could be playing an even bigger role in the rivers and forests of the Pacific Northwest. There was some indication that resident animals, from birds to rodents and insects, could help shape the forests as they fed in the understory, spreading nutrients and seeds, but no one had looked to see if nitrogen from the ocean might be playing a role in the forest as well.

Could he find salmon in the trees?

After moving to the University of Washington in 1987, Naiman got a small seed grant from the U.S. Forest Service that allowed him to collect some isotope data from the foliage of several streamside plants, including Sitka spruce, red alder, and ferns. Stable isotopes are the nonradioactive forms of elements, and our showrunner, nitrogen, comes in two forms: N^{14} and N^{15}. N^{14}, light nitrogen, is by far the most common nitrogen isotope. But the percentage of N^{15}, or heavy nitrogen, found in a given fish, bear poop, or leaf provides a chemical signature that reveals the source of the nitrogen. The carbon isotopes C^{12} and C^{13} can tell a supporting story, and isotopes of the two elements are typically examined together to understand the origins of animals, food, or nutrients.

Naiman and one of his students, Jim Helfield, set up their first study of marine-derived nutrients in the riparian plants on the Tenakee Inlet on Chichagof Island in southeast Alaska, an inlet fed by the watersheds of the Kadashan and Indian Rivers. They compared the growth rate and isotopic signatures of streamside plants in areas with spawning salmon to that of plants growing above waterfalls that blocked salmon migration. "Lo and behold," Naiman said, "the trees growing along the salmon streams were growing three times faster than those without salmon." They found salmon-derived nutrients in spruce needles along those rivers but not in needles near the rivers without salmon.

About a quarter of the nitrogen in the spruce needles and willow and poplar leaves of riparian forests appeared to be marine-derived. The trees absorbed the salmon-derived nitrogen and transported it up to their needles and leaves. As a result, they grew faster and taller—which was good for the salmon, as more shade and large woody debris provided cooler summer temperatures and river structure that aided in salmon reproduction and growth.

Nitrogen is critical to primary production, as we've seen with grasslands and phytoplankton. Without it, tree growth, soil fertility, and carbon storage are limited. The salmon-derived nitrogen was higher in the trees, devil's club shrubs, and ferns along the salmon runs. Only the red alders looked the same. These nitrogen-fixing trees can access nitrogen in the air and use it to help glean other elements from the bedrock they grow on. Animals can move organic nitrogen, but only microbes, some affiliated with plants like alders, can absorb it from the atmosphere and make it available to living organisms.

How do those nutrients get so far from the stream? Once salmon have reached their prime, they stop feeding, swim back to their native rivers, and then swim upstream to their natal spawning grounds—sometimes across hundreds and even thousands of miles. The lucky ones spawn before they die, and their bodies fall apart quite quickly.

Follow the body. In some cases, floods carry salmon carcasses from spawning streams to floodplains. Animals can move nutrients too. More than a hundred species that live along coastal rivers feed on dead salmon. Gulls peck at the eyes. Turkey vultures, bald eagles, ravens, and other scavengers feast along the riverbanks. Roughly 90 percent of the salmon biomass is consumed. The scavengers take wing, roost in the trees, and distribute nitrogen and phosphorus in the forest when they poop. Traces of salmon-derived nitrogen have been found more than a thousand feet from streams in Alaska, where much of this research has been done.

Bears eat salmon, some of which are captured at the metaphorical one-yard line, just before they spawn. Does a bear poop in the woods? Sometimes. Some brown bears stay close to the banks, along the river's edge; others are explorers, roaming far into the forest. In areas with lots of salmon, the biggest bears often stay put, defending their fishing holes, while the subdominant ones catch fish and move on, presumably to avoid getting in a tussle they're likely to lose. They might be responsible for carrying this marine-derived nitrogen deep into the woods.

Grant Hilderbrand, a biologist at the National Park Service, and colleagues found evidence that much of the nitrogen in white spruce that lined the rivers of the Kenai Peninsula, just south of Anchorage in Alaska, was delivered by brown bears. (And the loss of salmon in the Lower Forty-Eight is probably the worst thing to happen to grizzlies since the bear rifle.) More important, at least for our purposes, if a bear of any size poops in the woods, it's probably going to pee there too. "There's very little in the way of nitrogen in the poop," Naiman told me, "but there is a lot of salmon nitrogen in the urine." In fact, Hilderbrand estimated that about 96 percent of the nitrogen that bears distribute in forests comes through urine. Tom Quinn, a salmon biologist at the University of Washington, and colleagues estimated that bears moved roughly the same amount of nitrogen into riparian forests as a typical forestry operation would apply in a managed forest.

For twenty years, Quinn, Helfield, and their fellow researchers roamed a two-kilometer stretch of Hansen Creek in Bristol Bay, Alaska, moving every sockeye salmon carcass they found—217,055 of them, many killed by bears—on the right bank of the creek to the left. They found higher levels of oceanic nitrogen in the needles of the white spruce trees on the left bank. The carcasses delivered pulses to the forest soil, reaching the leaves within twenty years. That might be slow by our standards, but it's relatively quick for a tree in southwestern Alaska, at a latitude of almost 60 degrees north. They also found higher growth rates on the salmon-enhanced side. "Nutrients from salmon carcasses in proper abundances affect not only the isotopic signature but also the growth rates," Helfield noted. "They can be important for trees."

So why does this matter? The salmon life cycle and the massive pulse of nutrients the fish deliver are crucial aspects of forest ecosystems. The trees, streams, and salmon are all connected. Helfield and Naiman noted that riparian forests affect the quality of stream habitat by providing shade, filtering sediments and nutrients, and producing large woody debris. Salmon-borne nutrients improve the habitat for future generations of salmon and the long-term productivity of river corridors of the Pacific Northwest and Alaska. There is some evidence that this knowledge has long been encoded in local stewardship. After some tribes of the Pacific Northwest, such as the Nuu-Chah-Nulth, Haida, and Tlingit, smoked, dried, or cooked salmon, they buried the guts in the forest floor, fertilized blueberry bushes, and returned the bones to the streams to nourish the ecosystem, according to Suzanne Simard, author of *Finding the Mother Tree*. In a forest devoid of salmon, bears, and other animals, humans might have to labor to accomplish what happens in areas that are animal-rich. Living salmon are shot over dams by cannons to assist their migration. Volunteers distribute dead salmon or salmon analogues (fish pellets) in creeks in an attempt to restore traditional nutrient pathways.

Naiman retired in 2012, leaving a few things unfinished. "Each salmon carcass has about three thousand maggots, and there would be thousands of carcasses along the rivers," he told me over the phone. When the blowflies emerge, they spread through the forest, potentially moving nutrients to the forest floor through their frass—insect poop—and carcasses. They live only about three weeks from egg to adult. Nutrients can also move belowground—through tree roots and fungal networks that transport nitrogen and phosphorus between trees. "We were finding nutrients sometimes a hundred, two hundred meters away from the river. Much more than we could explain just by bear urine going everywhere." It looked like the trees were transferring the nutrients among themselves.

"Did you get what you need?" Naiman asked before we hung up. He was heading to Alaska the following week to go fishing. The coho were coming into Yakutat. What would he do with the fish? I asked. "I'm not a big fan of catching and killing a lot of them." Most of the salmon he caught would be released back into the river, where the bears, eagles, and perhaps the trees were waiting.

On my first full day at Nerka, the University of Washington's salmon research station in Alaska, upstream from Bristol Bay, Daniel Schindler, fisheries and aquatic sciences professor, took me by skiff to Pick Creek. He had agreed to show me what many ecologists consider the promised land of salmon runs. More than three and a half million sockeye returned to the region's streams each day I was there. Pick Creek was approaching peak salmon.

Clambering out of the aluminum boat in our waders, we saw a shimmering wave of salmon cresting at the mouth of the creek. Sockeye are shockingly bright—at times, the creek was so full of them, they looked like brake lights in a Friday-afternoon rush hour. Pick Creek and the area around Nerka is prime salmon-spawning habitat, with clear streams running over glacial gravel, huge protected areas, and tall green mountains. Much of the value here is in the heterogeneity, Schindler said. The watersheds of Bristol Bay form a vast mosaic of habitats—lake beaches, river valleys, glacial gravel beds, streams, and tea-stained wetlands—essential for the fish and other wildlife that live here.

Schindler had the lean, lined face of a man who had spent the past twenty-six summers in the Alaska sun and rain. He'd seen floods, droughts, heat waves, and cold spells run through these streams. This mosaic provided a portfolio effect, much like in investments, generating diversity and resilience in the face of perturbations and shocks.

For a moment, we were shin-deep in spawning salmon, a bunch of red eggs floating downstream. I bent down to touch the hump of a sockeye a couple of feet away, but it quickly swam off, eluding my reach. "Those fish have never seen a terrestrial predator," Schindler noted, "but their genes have."

We had motored past some fishers casting for freshwater rainbow trout earlier in the day. "The ocean comes to the fish here," Schindler said of the trout. "Those big rainbows are probably ninety percent marine-derived energy." They don't have to go to the ocean for seafood. Just about all of their energy comes through sockeye—the big, toothy males with pronounced humps and the smaller, sleeker females filled with eggs. By bringing the ocean to the rainbow trout, the salmon support a billion-dollar trophy fishery. The trout feed on the salmon eggs, and the float planes keep on coming, sometimes leaving with fillets, always leaving with stories.

Schindler tossed a couple of stranded sockeye back into the creek. These fish had beaten the odds; why not give them one more chance? The salmon we followed were the lucky ones, having survived a couple of years as juveniles, small fry, in the freshwater stream, then making it through smolthood—sockeye adolescence, when they move out to the sea—and finally swimming through a gauntlet of gill nets and fishing gear to get back here. Bristol Bay produces almost half of the global sockeye supply.

It's not just fishers who owe salmon a debt of gratitude; sockeye play a key role in sustaining the ecosystem. The circulatory-system metaphor seems quite literal here, with the streams pulsing like blood with bright red fish. "Life itself," quipped Tom Waits, "is really just the dead on vacation." (Judging by the struggles of these salmon, life is no party.) They were all short-timers around here: fast, spawn, die. The salmon ripple the streambed to lay their eggs among the gravel, disturbing the bottom—bioturbation. I tripped over one spawning female, then another. It was easy to imagine a conveyor belt, like the revolving sushi bars common in Japan, moving from the ocean to the mountains.

Schindler reached down and took the temperature of the water: 5.5 degrees centigrade (42 degrees Fahrenheit).

We rounded a corner, and there were thirty fish shoaling up in the gravelly shallows, getting ready to spawn. Later, when I helped Schindler and colleagues tag a couple hundred salmon in a nearby creek, I almost resented their abundance; each bend in the creek was another wall of fish.

"Hey, bear!" Schindler called out as we hiked up the stream. A surprised bear is a dangerous bear, so it's best to let them know you're coming, especially around a blind bend.

Judging by all the dead salmon with bite marks on their heads, bears were everywhere, trashing the salmon as they rode the red wave. By one account, brown bears near salmon streams occur in densities that are eighty times higher than in areas where salmon are inaccessible. At every turn in the creek, there were small tucked-away places where bears had hauled out salmon and pounded down the vegetation. Schindler called them bear kitchens—a "greaseball of dead salmon, bear piss, and mud." In the middle of one kitchen, there was a large bear print with a deep indentation of claws.

I felt for the bear spray clipped to my belt. "Does it work?" I asked.

"I think the stats are something like fifty-fifty that if you have a gun the bear goes away, and with pepper spray, it's about ninety-five percent," Schindler replied.

The can was still holstered there, to my relief.

"There's a paper in the *Journal of Wildlife Biology* by a Canadian who is one of the world's experts on bear attacks. There was one instance where a grizzly bear dragged someone out of a tent and ate him, so that was a predatory bear. But people get close to bears all the time—bear never touches 'em."

Still, that night, back at the research station, I was comforted by the can of bear spray hanging in the outhouse just to the left of the door latch.

"Hey, bear!"

At this point in midsummer, with the salmon still plentiful, the bears were super-selective, eating the fatty humps of the males right behind the eyes and clearing the eggs out of the females. That changes after the fish spawn. "By the end of the season," Schindler said, "they're eating skins and bones and just the junk that's left over." A typical run for a stream might be about three weeks; after that, the bears move on. "Time is the issue, not abundance of salmon," Schindler said. The bears have about three months to eat a year's worth of food.

Brown bears, like humpback whales, are capital breeders, storing up energy in the summer to get them through the winter. In June, a large male might be seven hundred and fifty pounds or so; it can gain four pounds every twenty-four hours, eating dozens of sockeye each day. By the end of the salmon run, the bear might weigh more than a thousand pounds. Katmai National Park, roughly a hundred miles to the east of Nerka, celebrates the gain with Fat Bear Week, a tournament where people judge which bear gained the most weight based on before and after photos. Sure, it's a gimmick, but it's a charming one, attracting more than six hundred thousand votes in October 2021. The winner was bear 480, or Otis, a four-time champ. The next year, bear 747 (aka Bear Force One) won for the second time.

Big bears like Otis and Bear Force One can increase their body mass by up to 30 percent. That's like a one-hundred-fifty-pound human gaining forty-five pounds in just a couple of months. Giacometti to Botero. Bears get obese. They get huge. And then they go into hibernation and disappear, surviving on their fat for up to half the year. Proteins, rich in nitrogen and phosphorus, help regulate their bodies through the winter. Their insulin levels don't change.

Remarkably, bears stay healthy as their weight goes up and down, retaining their strength and muscle mass.

As I watched salmon swim upstream and examined their bear-bitten carcasses, I had mixed feelings. These fish had come so close. Imagine surviving a few years at sea, a long, perilous migration (including the gill-net gauntlet that takes three of every five fish out of the breeding pool), and several weeks of starvation only to end up crushed to death in a bear's jaws. There's something almost tragic about the salmon that die at the end of this epic journey, just before spawning. At the same time, they die in their prime: still vigorous, with a purpose. And their end is swift, even if it comes just short of the goal line for many.

But their deaths are not in vain, since salmon and their nutrients support the bears, scavengers, and forest habitat for future generations of salmon and local wildlife. Quinn showed that about half the salmon eaten by bears had been carried into the forest. I looked at the trees that lined the creek and blanketed the mountains and wondered how much of the nutrients sustaining them had come from the ocean.

At the end of Pick Creek, we met a wall of forest. White spruce, paper birch, black cottonwood. A distinct end-of-road. To our right, there was a clear, flat stretch of spring soil under a spruce tree offering a view over the creek and the meadow to the south. A bear bedroom, Schindler said, with a kitchen and well-stocked pantry down the hill. It would've been a million-dollar listing on Bear Zillow.

PAT WALSH, MANAGER of the Togiak National Wildlife Refuge, came out to visit Nerka while I was there. "It's impossible for me to believe that there have been no changes to the ecosystem when fifty or sixty percent of the fish that would have returned have suddenly been taken away, and all the nutrients that this system evolved with have been gone now for a hundred and fifty years. Are we on the same path as Scotland, England, New England, Oregon, and

California, but just a little more delayed?" he said. His dog, who'd spent the past half hour chasing a stick into the lake, whimpered beneath the picnic table. "The reason the salmon are healthier here is because the land is still intact."

When you're at Nerka, it does feel like you've stepped back in time. Early colonists of North America described being able to walk across rivers on the backs of fish. You can't do that in most of North America anymore. But you almost could in the 1500s. "Europeans exploring Nova Scotia simply dropped baskets in near-shore waters and hauled up large cod," John Waldman, a biologist at Queens College in New York, wrote in the *New York Times*. In the Caribbean, Spanish sailors saw turtles that covered the sea. Whales were so numerous that they were "impossible to be counted." River herring once ran up rivers from the sea to spawn in quantities that were indescribable.

Around Nerka, there's no evidence of logging. The mountains are pristine and the wildlife abundant, despite the heavy fishing and bear hunting. I saw at least one bear every day when I was there, but most of the time they were running away: a mom and two cubs racing up the ridge, a bear slinking back into the bush after spotting us motoring by. We mostly saw their prints and muddy brown scats.

Before coming to Alaska, Walsh had been a biologist for the Department of Defense at Avon Park Air Force Range, where he oversaw twelve endangered species. "It was really interesting work, but watching the grasshopper sparrow get less and less each year, the Florida scrub jay go down, and the red-cockaded woodpecker decline—it wears on you."

George Pess, a former Schindler student who works on salmon restoration at the Northwest Fisheries Science Center in Washington, said that it sounded a lot like working on salmon in the Lower Forty-Eight. The range of Pacific salmon, six species in the genus *Oncorhynchus* that are all semelparous (defined by a singular reproductive burst at the end of their lives), extends from Japan to California. In the Pacific Northwest, the declines have been catastrophic, from about seventy million fish before industrial dams and commercial fishing, roughly a hundred years ago, down to just five million now. For every hundred salmon that swam the rivers of the region a century ago, a mere six remain. Twenty-nine salmon populations are protected by the Endangered Species Act. Many of the remaining salmon are raised in hatcheries, often for human consumption. Others are shipped around the enormous dams on barges or in trucks. "On I-84," Lewis and Clark professor Dan Rohlf said, "you can get passed by migrating salmon."

The Columbia River, with headwaters in British Columbia, runs through seven states, defining much of the border between Washington and Oregon before it hits the Pacific Ocean. A century ago, it might have held as many as fifteen million salmon spawning each year, including three types of Chinook or king salmon, coho, sockeye, steelhead, and chum. Decades of overfishing and habitat destruction—irrigation, logging, and grazing—have taken their toll. So has the Grand Coulee Dam west of Spokane, which stands almost forty stories high and a mile wide and cuts off the natural migration of fish on the Upper Columbia. Now, fewer than two million salmon migrate in the river each year. In British Columbia, more than thirty million sockeye once ran the Fraser River in four-year cycles. But that was before humans started messing with the rivers and the fish. In 2020, there were fewer than four hundred thousand.

Brown bears, wolves, and other hunters and scavengers have lost an essential food source. There were probably more than fifty thousand grizzlies ranging over a million square miles of the West when Europeans arrived. Now there are just fifteen hundred in a tiny fraction of their former range, mostly around Yellowstone. The loss of the fish goes far beyond the bears; it ripples out through the forests and waterways. Forests can lose nitrogen, phosphorus, and other nutrients, slowing growth along the riverbanks and leaving waterways vulnerable to erosion. Streams run out of nutrients and algae disappear. The scientific term for the condition is *oligotrophic*—a deficiency in nutrients that can be bad for the forests and water-dwelling invertebrates. It's also bad for the salmon. In the absence of parental carcasses, young-of-the-year salmon lose weight, grow more slowly, and have lower genetic diversity. Fewer survive.

"DON'T HAVE A heart attack here," Schindler called over his shoulder as we were hiking up Church Mountain, out behind the research station. This wasn't anything like the old switchbacks I was used to back east in Vermont. The trail, when there was one, went straight up to the peak. I was breathing so hard, I could barely talk. Now my pulse was racing even faster.

One in our party had already dropped out, deciding to take his chances with the brown bears rather than struggle uphill at Schindler's breakneck pace.

"Sometimes you'll find the salmon carcasses halfway up a mountain, and you'll think, *Holy shit, that's a lot of nutrients,*" Schindler said, "but when you add it all up, it's a tiny fraction of the population."

"So what about the bear pee and poop?" I asked him. "Do you think they bring nutrients out of that watershed into the forest?"

"They do, but it's only within a couple meters." Forests being fertilized by carcasses and bear pee is "total bull," he told me. "I tell most people that it is a story just like the Bible—too good to be true, and the facts are rarely checked." During several discussions, Schindler expressed his concern that much of the work relied on stable isotopes that were confounded by watershed features. Floodplains have water-saturated soil that can elevate signals associated with marine-derived nutrients. Never one to pull his punches, Schindler objected to how the tree measurements had been taken and questioned whether nitrogen had much effect on productivity in the region anyway.

Schindler had written a lot over the years about the seasonal pulses of nutrients tied to the salmon migration. Salmon lay their eggs in the interstitial spaces between the gravel. I could feel this disturbance as we walked up Pick Creek, where the creek bottom was often unstable, in the wake of salmon building their redds, or spawning nests. His work showed that salmon brought more nitrogen and phosphorus in their bodies up from the ocean than they released downstream, but they also churned up a lot of sediments that got released into the currents.

We sat on a windy bluff near the top of Church Mountain. Lake Nerka stretched below us, hemmed in by green mountains. One experiment Schindler and his team did was in bear kitchens, measuring the nitrogen in the soil. "When the bears are active, nitrogen is cranking," Schindler said, but when they put up electric fences to keep the bears away, everything went back to baseline after a year. The microbes thrived. They're essential to a healthy ecosystem, too, of course, but they weren't fueling the growth of the nearby plants.

Schindler didn't deny that animals were important: "To pretend that animals have no effect on ecosystem processes is a joke, yet so much of biogeochemistry is unwilling to embrace that." Certainly, many animals benefited from the migration—the fish that eat salmon eggs, the bears that eat salmon and other fish. "The whole scavenging community is huge," he said. For a short period, in a concentrated space: "Things go nuts. The rate at which these carcasses disappear is actually hard to believe."

I asked Gordon Holtgrieve, a professor at the University of Washington and one of Schindler's former students who was visiting Nerka at the time, about the salmon-derived nutrients. "There's really convincing evidence that dozens, if not hundreds, of species in these systems are completely dependent on salmon for growth," he said. "The consumer-level effects are totally clear. It's the nutrients-as-fertilizer idea that's more shaky."

He didn't challenge the data, but he felt there were some very rosy interpretations of the salmon effects. Holtgrieve acknowledged that his was a small voice in the wilderness. "It's a funny debate," he said. "The anti side is so small, it's not really a debate. But we had pretty different conclusions." He had a point. I had found this pocket of skepticism—or it had found me—by chance. Schindler had invited me to give a talk on whale ecology at the University of Washington, and when I mentioned my interest in the bear-salmon interactions, he dropped what I think of now as the Schindler bomb: "It is mostly a myth! Seriously...there has been a lot of poorly executed science and vastly overinterpreted stable-isotope data to construct this 'story.'"

"We have a summer place on the other side of the mountains," Holtgrieve said, "and it's got a small Chinook run in the river. My daughter was seven or something. She found a spine and one gill plate of a Chinook. A tiny thing. She brings it back to me and says, 'Look what I found.'" He told his daughter that it was a Chinook, or king, spine, showed her how to identify it, and later told a local biologist about the find.

What had he done with the spine?

"Oh, I threw it in the compost bin."

The biologist told him that the Washington Department of Fish and Wildlife would be really mad at him for doing that.

"Why?"

"You took the marine-derived nutrients out."

"They're missing the forest for the trees in this case," Holtgrieve told me, "because if I go down and take a leak, I'll replace the nutrients."

ON MY LAST full day at Nerka, I joined Schindler and George Pess for a sockeye survey on Allah Creek. Pess counted the dead fish, Schindler the living. You couldn't miss the live ones, vivid red and determined to make their way upstream. A few were close to the end, out of energy and losing their bright colors. The dead were like dirty gray dishrags, draped over branches or swirling in eddies.

We walked through a tunnel of alder, cottonwood, and white spruce with occasional flashes of fireweed. Stopping every two hundred yards or so, Schindler tallied the numbers in precise handwriting in his Rite in the Rain notebook. The ratio of living to dead was about two to one.

I watched two salmon mating on what might have been the last day of their lives. They were at the end of their epic journey. After two or three years at sea, they had gone without food, fought currents, hurdled small waterfalls, and swum many miles upstream. Elsewhere, sockeye run nine hundred miles and sixty-five hundred feet in elevation.

It's still not quite clear how salmon locate their natal streams. Chinook salmon, a close relative of the sockeye, provide a hint. They appear to use Earth's magnetic field to find the entry to the rivers. From there, it's believed that the salmon's memories of smells unique to certain soils and vegetation help guide them to their natal streams.

Eat, fast, spawn, die—in that order, if they're lucky. As the columnist Maureen Dowd put it, "The timing of your exit can determine your place in the history books." In this case, the timing of each salmon's exit determined its place in Schindler's neatly written field notebook.

Schindler was as determined as the salmon to make his way upstream. The gravel was slick with algae, and Pess and I struggled to keep up, grateful for the hiking sticks Schindler had handed us when we got out of the boat. Every hundred yards or so, we passed a bear kitchen, some a slick mess, scattered with fishbones and eggs, and others just a broken path through the underbrush. Near one of the kitchens, we saw bright eggs atop the moss and a headless carcass in the mud, toothmarks in its back, covered in maggots. The carcasses were now fair game for flies and other insects.

"Hey, bear!"

We stopped for lunch on a gravel bar close to a bear kitchen. Schindler scooped some herring out of a tin with his crackers. He rinsed the can in the river, tucked it into his backpack, and suggested taking a shortcut back. "It's a bit of a scramble."

Pess and I looked up at a wall of brown scree. Perfect for a grizzly, maybe. But us?

Schindler got a foothold in what looked like the pawprints of a bear and started clawing his way up the hill. A writer from *Science* once described Schindler as having an uncanny resemblance to a grizzly, and I could see it here: the muscular shoulders, heavy brow over close-set eyes, and ursine brown hair poking out from beneath his baseball cap. From our angle, the slope looked impossibly steep, maybe thirty degrees. Pess and I watched as the scree started falling harder and faster. The earth was shaking, and it seemed like half the mountain would come down around us. The falling stones started to get bigger. After one almost hit me in the chest, Pess and I stepped back into the creek.

Schindler peered over the ridge. My turn.

Any sign of bear prints or easy footholds in the chute had been clawed away, so I opted to clamber up along the alders from branch to branch, digging my boots into the scree. Once I got near the top, I still had to make it across the open slope to the other side, where Schindler was waiting.

I hesitated, then took the plunge, digging into one of Schindler's prints or perhaps a bear's. I clutched at the mossy bank and belly-flopped onto the ridge like an ancient pteropod testing out land for the first time.

Below us, the alders were shaking. It looked like a bear was heading toward us, but it was Pess, clutching his way through the branches with all the determination of a beleaguered salmon struggling to make it over a final wood-filled waterfall. He lost his footing as the scree in the chute cascaded down. Schindler reached out, grabbed the end of Pess's hiking stick, and guided him to the bluff.

"That's why I went first," Schindler teased.

Pess and I followed him over the hummocks, long grasses obscuring the turf. I had recently sprained my ankle, and I couldn't help thinking that an injury here would be a nightmare. The thought of spending the night among the ferns and birches waiting for a rescue mission was not appealing, not even with my can of bear spray. Our trek seemed to go on for hours.

"How much longer is this shortcut?" I asked.

We passed through a broad meadow with a bear wallow and then slipped down a muddy stream before returning to the main creek again. It looked like we were done for the afternoon—sure, it had been a lot of effort for thirty data points, but that's often the nature of fieldwork.

Schindler had other ideas. On our way down the Allah, he said, we would collect 220 half-rotten fish and remove their tiny otoliths for later analysis. The "earstones" would reveal whether the salmon were two- or three-ocean fish (meaning whether they'd spent two years or three at sea) and where they had spawned.

"Do you want to collect them," Schindler asked, sharp knife in hand, "or would that offend your East Coast sensibilities?"

I worked my way down the creek, lining up the dead bodies on gravel bars and in bear kitchens.

"Bring out your dead," Pess called as he hacked into the heads and extracted the otoliths.

In this *danse macabre*, the salmon were in various states of decay. Many had clearly passed over, their red bodies now gray. Gulls had pecked the eyes out of a few of them. Others looked like they were on the edge of death—one male, still bright red, haunted me in this idyllic land: it moved its jaws while I lined it up with the others. The most far gone, the maggot-infested, were called "crackers" and largely avoided. We sampled 110 females and 110 males. By the time we got back to the boat, I was pretty sure I never wanted to see a dead salmon again.

Later that night, a student from Japan made a beautiful plate of salmon sashimi that briefly stopped our conversation. Holtgrieve grilled sockeye by the lake. They tasted pretty good.

WHEN CONSTRUCTION BEGAN in 1910, the Elwha Dam was designed to attract economic development to the Olympic Peninsula in Washington, supplying the growing community of Port Angeles with electric power. It was one of the first high-head dams in the region, with water moving more than a hundred yards from the reservoir to the river below. Before the dam was built, the river hosted ten anadromous fish runs. All five species of Pacific salmon—pink, chum, sockeye, Chinook, and coho—were found in the river, along with bull trout and steelhead. In a good year, hundreds of thousands of salmon ascended the Elwha to spawn. But the contractors never finished the promised fish ladders. As a result, the Elwha cut off most of the watershed from the ocean and 90 percent of migratory salmon habitat.

Thousands of dams block the rivers of the world, decimating fish populations and clogging nutrient arteries from sea to mountain spring. Some have fish ladders. Others ship fish across concrete walls. Many act as permanent barriers to migration for thousands of species.

By the 1980s, there was growing concern about the effect of the Elwha on native salmon. Populations had declined by 95 percent, devastating local wildlife and Indigenous communities. River salmon are essential to the culture and economy of the Lower Elwha Klallam Tribe. In 1986, the tribe filed a motion through the Federal Energy Regulatory Commission to stop the relicensing of the Elwha Dam and the Glines Canyon Dam, an upstream impoundment that was even taller than the Elwha. By blocking salmon migration, the dams violated the 1855 Treaty of Point No Point, in which the Klallam ceded a vast amount of the Olympic Peninsula on the stipulation that they and all their descendants would have "the right of taking fish at usual and accustomed grounds." The tribe partnered with environmental groups, including the Sierra Club and the Seattle Audubon Society, to pressure local and federal officials to remove the dams. In 1992, Congress passed the Elwha River Ecosystem and Fisheries Restoration Act, which authorized the dismantling of the Elwha and Glines Canyon Dams.

The demolition of the Elwha Dam was the largest dam-removal project in history; it cost $350 million and took about three years. Beginning in September 2011, coffer dams shunted water to one side as the Elwha Dam was decommissioned and destroyed. The Glines Canyon was more challenging. According to Pess, a "glorified jackhammer on a floating barge" was required to dismantle the two-hundred-foot impoundment. The barge didn't work when the water got low, so new equipment was helicoptered in. By 2014, most of the dam had come down, but rockfall still blocked fish passage. It took another year of moving rocks and concrete before the fish had full access to the river.

The response of the fish was quick, satisfying, and sometimes surprising. Elwha River bull trout, landlocked for more than a century, started swimming back to the ocean. The Chinook salmon in the watershed increased from an average of about two thousand to four thousand. Many of the Chinook were descendants of hatchery fish, Pess told me over dinner at Nerka. "If ninety percent of your population prior to dam removal is from a hatchery, you can't just assume that a totally natural population will show up right away." Steelhead trout, which had been down to a few hundred, now numbered more than two thousand.

Within a few years, a larger mix of wild and local hatchery fish had moved back to the Elwha watershed. And the surrounding wildlife responded too. The American dipper, a river bird, fed on salmon eggs and insects infused with the new marine-derived nutrients. Their survival rates went up, and the females who had access to fish became healthier than those without. They started having multiple broods and didn't have to travel so far for their food, a return, perhaps, to how life was before the dam. A study in nearby British Columbia showed that songbird abundance and diversity increased with the number of salmon. They weren't eating the fish—in fact, they weren't even present during salmon migration. But they were benefiting from the increase in insects and other invertebrates.

Just as exciting, the removal of the dams rekindled migratory patterns that had gone dormant. Pacific lamprey started traveling up the river to breed. Bull trout that had spent generations in the reservoir above the dam began migrating out to sea. Rainbow trout swam up and down the river for the first time in decades. Over the years, the river started to look almost natural as the sediments that had built up behind the dams washed downstream.

The success on the Elwha could be the start of something big, encouraging the removal of other aging dams. There are plans to remove the Enloe Dam, a fifty-four-foot concrete wall in northern Washington, which would open up two hundred miles of river habitat for steelhead and Chinook salmon. Critically endangered killer whales, downstream off the coast of the Pacific Northwest, would benefit from this boost in salmon, and as there are only seventy individuals remaining, they need every fish they can get.

The spring Chinook salmon run on the Klamath River in Northern California is down 98 percent since eight dams were constructed in the twentieth century. Coho salmon have also been in steep decline. In the next few years, four dams are scheduled to come down with the goal of restoring salmon migration. Farther north, the Snake River dams could be breached to save the endangered salmon of Washington State. If that happens, historic numbers of salmon could come back—along with the many species that depended on the energy and nutrients they carry upstream.

Other dams are going up in the West—dams of sticks and stones and mud. Beaver dams help salmon by creating new slow-water habitats, critical for juvenile salmon. In Washington, beaver ponds cool the streams, making them more productive for salmon. In Alaska, the ponds are warmer, and the salmon use them to help metabolize what they eat. Unlike the enormous concrete impoundments, designed for stability, beaver dams are dynamic, heterogeneous landscapes that salmon can easily travel through. Beavers eat, they build dams, they poop, they move on. We humans might want things to be stable, but Earth and its creatures are dynamic.

***

ON THE OTHER side of the continent, antigravity forces are propelled by reproduction of another kind. In spring, millions of sea turtles come out of the ocean to lay their eggs.

Karen Bjorndal, director of the Archie Carr Center for Sea Turtle Research at the University of Florida, has spent much of her life thinking about marine reptiles. She first observed them in the Galápagos during a summer fellowship when she was in college; she lived for three months on the island of Santa Fe, which had no other human inhabitants. "I was there by myself camping in a little tent studying the social behavior of land iguanas," she told me, "but I quickly discovered that the land iguanas didn't do very much." Since she didn't have anyone to keep her company, she spent her spare time staring out over the sea, watching the sea turtles swimming past and coming up to blow. "I was intrigued by the feeling that they're in two different worlds," she said—they bridged that gap between sea and air.

Bjorndal soon applied to work with the legendary sea-turtle biologist Archie Carr at the University of Florida. She and her colleagues spent years on nesting beaches, from Tortuguero, Costa Rica, to the Bahamas, tracking the mostly positive population trends of many sea turtles after catastrophic declines in previous decades. They watched hundreds, and in some cases thousands, of turtles emerge from the sea at night to lay their eggs. After about eight or ten weeks of observation, Bjorndal would dig up the nests and count the eggs and the shells to examine nesting success. How many nestlings had emerged; how many had died in the sand? She became entranced by how the sea turtles could transform seagrasses, which have few grazers, into these "nitrogen-enriched particles" otherwise known as eggs.

"It's an extremely stinky job," she said. The nests smell like, you guessed it, rotten eggs. "But going through all this biomass, you can't help but think, *Wow, there's a lot left over here.*" Even the nests from which the greatest number of hatchlings made it to the sea were filled with goo.

Her suspicion that sea-turtle nesting behavior represented a nutrient transfer from sea to land sat on the back burner until she got a faculty position at the University of Florida. There, she and Sarah Bouchard, a master's student, set out to measure the amount of nitrogen and lipids deposited in loggerhead sea-turtle nests in Florida. Their ultimate goal was to see how much of the nutrients went back into the ocean and how much remained in the vicinity of the nest. "We were surprised to find that only about a third of all of those nutrients returned to the sea as hatchlings," Bjorndal told me. Tens of thousands of green and loggerhead sea turtles migrate from rich foraging grounds to nutrient-poor beaches in Florida each year. Sea-turtle nests could be a crucial source of energy, fats, nitrogen, and phosphorus for these coastal ecosystems.

The Florida black bears surely loved them. Early naturalists wrote that they ignored the turtles but adored their eggs, which were easy to find and full of fat and nutrients. Bears are good diggers and great eaters, devouring in a day what most humans could eat in a week. As the bears moved around the landscape, they transported the marine nutrients even farther from the shore.

Another student followed up Bouchard's study using stable isotopes, showing that these nutrients from sea-turtle eggs were being absorbed by sea oats in the dunes. This had an impact on beach structure. By stabilizing the dunes, the nutrients deposited in nesting grounds help preserve the habitat for future generations of sea turtles—which, like salmon and whales, have a high degree of fidelity to the place where they hatched or were raised. It also helps humans. We like reliable dunes, especially in times of hurricanes or rising sea levels.

ONE AFTERNOON IN Nerka, I sat by a creek watching the water roll by. There were hundreds of salmon, some caught just below a small waterfall. Occasionally one found a path upstream. I could see the salmon falling apart before my eyes, their bright skin peeling off into the stream. "In the moment before death," neurobiologist David Eagleman wrote in *Sum*, "you are still composed of the same thousand trillion trillion atoms as in the moment after death—the only difference is that their neighborly network of social interactions has ground to a halt. At that moment, the atoms begin to drift apart, no longer enslaved to the goals of keeping up a human form."

Or, in this case, keeping up a salmon form. As fish biologist Gary Lamberti described it: "They've invested so much of their elements into their gametes and they've excreted so much that what's left is kind of a recalcitrant body." Most of the good stuff had already been released in the form of eggs and sperm and the metabolic waste associated with swimming upstream without any food.

In the months before and after my trip, I read dozens of papers about salmon migration, nutrient subsidies, and foliar N^{15}, the chemical signal that showed that nitrogen from salmon had gotten into the trees. The integrity of that story started to fray as Schindler and Holtgrieve pulled at some of its essential threads, but neither one of them denied the widespread effect of salmon on the animals that live in the forests of the North Pacific.

Salmon shape ecosystems in surprising ways. At least twenty species have been seen eating salmon, including eagles, ravens, crows, mink, marten, coyotes, and wolves. Blowflies, the most abundant consumers of salmon by sheer number, time their emergence to synchronize with the sockeye's return. This makes sense, given that blowflies, also known as carrion flies, get about 85 percent of their food from salmon carcasses. They are usually the first insects to colonize a salmon corpse, and one blowfly, with bright red eyes and a body gleaming blue like a new car, will lay about a hundred and fifty to two hundred eggs on a salmon carcass. They hatch about eight hours later. Once they emerge as adults, after about a week, they feed on flowers and serve as pollinators. Schindler and Peter Lisi, one of his grad students, found that the kneeling angelica, a common streamside flower that vaguely resembles Queen Anne's lace, also times its bloom to follow the salmon run and the blowfly emergence. More flies means more pollination and more seeds.

Brown bears depend on salmon throughout much of their range. In Alaska, these fish-fueled bears are also crucial seed spreaders. Bears are abundant and big eaters, pooping seeds that are primed to germinate. I had seen some bear scats, mud brown with red highlights, that I assumed were from salmon eggs. Turns out they were cranberries. They also spread blueberry seeds, up to 37,000 in a single scat, and thousands of devil's club, raspberry, and currant seeds through muskeg, forests, and other habitats. Small mammals such as voles eat the scat, moving the seeds even farther.

One often-cited study done in the Great Bear Rainforest in British Columbia found, perhaps counterintuitively, that streams with lots of salmon have reduced plant diversity. Wait—more salmon equals fewer species? Just like at Surtsey, the addition of high amounts of nutrients—whether from bird poop, dead salmon, or bear pee—can favor a few dominant species. In the case of the Canadian rainforest, the well-named salmon berry, a plant that thrives in high-nitrogen environments, outcompeted azaleas and blueberries in areas near high spawning densities.

In the Mokelumne River basin of California just north of San Francisco, the main riparian crops are wine grapes. Researchers found elevated levels of marine nitrogen in the grapevines near salmon streams. About a quarter of the nitrogen in the vineyards along the Mokelumne comes from the ocean. In this case, it's not bears moving it from the rivers to the soil. Brown bears are long gone from the region, hunted relentlessly in California until 1922, when the last one was killed in Tulare County. Nowadays, turkey vultures are the most common scavengers of salmon carcasses in the area. I tried to find a salmon-derived vintage that didn't blend with nearby vintners, but to no avail. If there was a salmon terroir—a *merroir?*—I couldn't unearth it.

***

I LONGED FOR the simplicity of Surtsey, where the arrival dates and abundance data of just about every species were well documented. Complex ecosystems like the Pacific Northwest, with thousands of species, are another story, with billions of interactions, direct and indirect, among species. A direct effect is relatively easy to observe. If a bear eats a salmon, the bear wins, the salmon loses. But indirect effects ripple out through the food web, across animals, plants, and fungi—a dead salmon affects the pollination of a streamside plant and the productivity of a tree.

The salmon, the bears, and the trees make a beautiful story, one that sits squarely in the framework of this book. One that is relatively tidy and clear. And while there is quite a bit of data to support it, there is also a strong case for skepticism. Ecology is messy, and the tools we use, such as light and heavy nitrogen, are sometimes crude and limited by funding, seasons, weather, and even pandemics. Scientific inquiry is, like just about any human endeavor, complex, susceptible to interpretation, even fraud, and at times infused with biases of all kinds. People search for an easy answer even if there isn't one.

New discoveries are often accompanied by a fury of excitement that is soon followed by debate and challenges as scientists from different disciplines and locations enter the fray, poking holes in the original idea. With time and research, a deeper understanding should emerge. Then new uncertainties will arise and the whole cycle continues. As neuroendocrinologist Robert Sapolsky said of his research on baboons: "The debate rages on, keeping primatologists off the dole."

When I mentioned Schindler's skepticism about salmon-derived nitrogen making it into trees to Naiman, he was surprised. "I haven't seen Daniel for a number of years," Naiman said, "and he thinks deeply about things. But he often will take the path less followed, and he'll ask hard questions." It was a warm response, and I immediately thought of Schindler clawing up the scree like a grizzly bear. Naiman pointed me to a study showing that the amount of marine-derived nutrients in trees went down after dams were constructed in Washington and Oregon. When the researchers cored the trees, they found that the growth rates of the trees returned to ambient levels within two years of dam construction. "They no longer had a subsidy from the salmon carcasses," Naiman explained. It wasn't a smoking gun, perhaps—and the data to me seemed a little messy—but it was concrete evidence supporting the numerous papers that examine salmon-derived nutrients.

When I told a friend in the Seattle area about the debates surrounding the salmon in the trees, she pushed back. The organic salmon fertilizer she'd been using in her garden, she told me, had done wonders for her vegetables. Just as the stories that recreational fishers carried back from Alaska were as important—perhaps more so—as the frozen meat they took home, I wondered if her belief in the enhanced value of the fertilizer was driven by the data, the beauty of the story, or some combination of the two.

"Ecology isn't rocket science," Schindler had quipped when we finished our hike on Church Mountain. "It's a lot harder."

4

Heartland

On a rainy Monday morning in mid-June, there was a knock on the door of my hotel room in Cooke City, not far from the northeast entrance to Yellowstone National Park. It was the manager of the Elk Horn Lodge.

"I don't know if you've been following what's going on, but the park is closed and the road to Silver Gate is flooded." I had spent the morning answering e-mails, looking out the small window at the downpour, and waiting for the weather to clear.

Five inches of rain had fallen since I had arrived the day before, melting the snow on the mountains and adding runoff to the rain. The town was expecting power outages and the loss of drinking water.

I took a walk around the rain-soaked town. It was pouring; a muddy stream running down the main road was up to my ankles. The water was up to the windows of some of the low-lying houses.

I wasn't ready to leave just yet. I had traveled for more than twenty-four hours to get to Yellowstone to accompany several bison biologists into the backcountry for a couple of days. I thought I might stick it out.

I ran into the owner of the lodge, who was also a first responder, and he told me that the water was up to the bridges on the road to Red Lodge, the only route out. The night before, there hadn't been an empty room in town; now the streets felt deserted. As we chatted, the last of the rental cars were leaving town. Rescue workers were moving in.

"You should go."

I WAS SITTING with canid biographer Rick McIntyre in the Lamar Valley when a black wolf walked out of the clouds. This was the day before the knock on the door, and I had been tempted to sleep in. The weather didn't look great, I'd just gotten into town, and I had five days scheduled in the park, including a night in the backcountry. Why rush? But I found myself getting up at 3:30 a.m. anyway and meeting him near his cabin at Silver Gate, close to the northeast entrance of Yellowstone.

Never say no to a day on the water or a morning in the field, especially if it's with McIntyre. I'm glad I didn't. We had no idea at the time that it would be the last regular day of the season.

I had visited McIntyre and his wolves more than a decade earlier, in 2008, to see one of the great successes—and controversies—of the Endangered Species Act. Gray wolves had been reintroduced to Yellowstone after an absence of almost seventy years. McIntyre said that it was peak wolf-watching that year, and the balance between the valley's largest herbivores, elk and bison, was starting to change.

This is how I recall the first day of that trip: The heart of the world lay open, rippling with sunlight. Rick McIntyre and I sat among elk bones overlooking the confluence of the Lamar River and Soda Butte Creek. It was June, and McIntyre was on a folding chair, his long legs outstretched beneath a tripod that held his spotting scope, the valley spread out in front of us like an open textbook. On the right bulged Specimen

Ridge, snow snaking down the northern valleys, with white-capped Mount Norris on the left. Big sage covered the hillsides like an old pilled sweater. Swollen with snowmelt, the Lamar ran steadily. Dead trees lay scattered along the bank among live cottonwoods and willows. Along the floodplain, buffalo moved in drifts. The cows were molting, their coats in tatters; on the brightness of the fresh grass, it was the young that stood out, a smoldering ocher. With their snowplow heads and Minotaur shoulders, they were improbably beautiful.

One afternoon on that first trip, I watched a herd of bison fight off the Druid Peak wolves, the local pack, who were eager to get at an orphaned calf. With its mother dead, the calf's death was predetermined at this point. Bison don't wet-nurse, so at best the calf would starve. But the battle was bravely fought, and as darkness fell, the calf was still standing. By the next day, it was a distant mound of fur and bones. Ravens, perhaps America's top scavengers, had moved in.

Back then, the Lamar Valley, in northeastern Yellowstone, was still pretty sleepy. People knew about the wolves, but there were few wolf-watching companies. Most of the tourism still revolved around Old Faithful and West Yellowstone.

Leaning against a boulder fourteen years later, McIntyre regaled me with stories about wolves he had known and the valor in wolves' lives. He described the last days of Wolf 911, an alpha male he had followed for years who came up against a rival pack. "The other wolves chased him down to the far bank," McIntyre said. "They stared at him from across the creek, eight against one. The smart thing for him to do would be to abandon his kill and cross the road and go up to where his family was and he'd be safe. But he didn't do that. He stood his ground, and they eventually swam across and surrounded him. He was giving them every indication that he had no problem fighting it out with them, even though he was so injured and so old. They attacked him, and he fought back well. But no one wolf can defeat eight opponents, and they killed him.

"I later thought, *Why didn't he just save himself?*" McIntyre continued. "Because he could have crossed the road and reunited with his family. But that night, those wolves could have followed his scent trail; it would have taken them right to his family, and that's where the attack would've been, as opposed to out here."

I think it is safe to say that McIntyre has spent more time observing wolves than any human in the history of the world. When I visited in 2022, his total was 9,020 days. "Not that I'm counting," he joked. He had reached 6,175 consecutive days before he had to stop for a heart operation.

"Why vacation?" he said, looking over the Lamar. "Where would I go?" We gazed out toward the Druid Peak wolf pack's den, and I had to agree; the park truly is one of the most beautiful places I have ever seen.

When I first met him, he'd told me he was planning to write a book about Yellowstone wolves, but at the time he was spending about nine hours a day in the park. I doubted that his meticulous notes would ever make it onto the printed page. How wrong I was. McIntyre has since written what I think of as the canine version of Durrell's Alexandria Quartet or Ferrante's Neapolitan novels: *The Rise of Wolf 8, The Reign of Wolf 21, The Redemption of Wolf 302,* and *The Alpha Female Wolf.* He recounted the battles and the blood feuds that developed between packs and the deep devotion among alpha wolves. With a new book coming out every year and the long hours he spends in the park, McIntyre may someday rival Balzac with his own *La Comédie du Loup.*

Soon after sunrise, a light rain began to fall. I was surprised by the number of campers, cars, and wolf-watching vans snaking along the Lamar River. It was no longer a sleepy corner of Yellowstone but a must-see stop when visiting the national park. Yellowstone is the only place in the United States where bison have lived continuously since they came perilously close to extinction in the nineteenth century. And since the wolves were restored, the park has most of the major players it had before people arrived: brown bears, mountain lions, elk, pronghorn, coyotes, beavers, wolves, and bison.

Yellowstone seems like a prime location to observe how large animals can build ecosystems.

The rain got heavier, soaking our boots as we crossed the grass. We paused at a bison patty. I reached down and broke off a piece. It was brown, like most mammal poop, a mixture of digested food—grass, mostly, in this case—and dead blood cells. It smelled like fresh soil with a slight whiff of cut grass. There were mushrooms growing out of one side. I had heard that the fungi were safe to eat, perhaps even hallucinogenic. Poop is patchy. It's ephemeral. Mushrooms need to breed quickly, and many require partners, often relatives. Dung-associated fungi can use psilocybin as a weapon to help them defend their fecal realm, drugging the competitors and predators, usually insects, that are drawn to the resources provided by bison and other animals.

I noticed more bison on this trip than I remembered seeing in 2008. They were grazing on Kentucky bluegrass, clover, and dandelion, revealing the agricultural history of the Lamar Valley. The wolves seemed less visible, perhaps, McIntyre noted, because the alpha females of the local pack were in a power struggle, so territories were fluid and uncertain.

We paused to read a new sign on the road explaining the wolf effect:

The reintroduction of wolves brought change to an ecosystem that evolved without them for almost 70 years. Although wolves do not directly affect all life around them, their effects possibly tumble down the entire food chain. This hypothesis is called a trophic cascade.

In the absence of wolves, the plaque continued, unnaturally large herds of elk swelled in the park. Once wolves returned, the herds got smaller and stronger. And wolves created a landscape of fear; elk lingered less along streams, and as a result, native willows, aspens, and cottonwoods grew taller. Birds found more places to nest in these riparian trees, and beavers returned. And the changes also benefited fish, which are attracted to the cooler, shaded waters. Likewise, newly formed beaver ponds were good for native waterfowl, amphibians, and reptiles.

I asked McIntyre about the wolf effect. Had he noticed these changes?

"There's definitely more successful regeneration of aspens and willows," McIntyre said. A beaver colony was established in Crystal Creek soon after the wolves were reintroduced to the area. And though the beavers moved upstream over time, there were still plenty of aspens and willows for food and dam construction. A new pack of wolves was denning in the area.

McIntyre got word that a wolf had been sighted near Soda Butte, so we packed up our gear. Even after more than nine thousand days tracking wolves in Yellowstone, he was always racing off to see the next one. His records are valuable because he has seen just about every Yellowstone wolf and reported every sighting, no matter how fleeting.

In contrast to the increased bison traffic, there seemed to be fewer elk than when I last visited. Was it because of the wolves? "It's hard to say," McIntyre noted. There was probably some selective predation: "Healthy elk and bison can take care of themselves, but wolves smell the fear or follow any that might take off running." Populations don't change much if predators go for the very young or elderly, which have lower reproductive rates than animals in their prime. But there had been more hunting pressure in Montana to the north of the park since 2008, which might have influenced elk populations. The elk and the ecosystem had changed since the wolves arrived, but the root causes of these changes were complicated.

According to Wyoming biologist Matt Kauffman, the reintroduction of wolves had little behavioral impact on elk. While he was working on his PhD, he received a grant to examine the hypothesis that wolves could scare elk out of places where trees were over-browsed, and the trees would recover. But after he began running the experiments, he started seeing all these holes. Wherever he and his colleagues protected aspen with fences, the trees thrived. But Kauffman saw little variation between the areas where wolves had killed elk and areas where they hadn't. "There was no difference between risky places and safe places," he said. He thinks this is partly because elk weren't living with wolves every day. "It turns out that an elk has to get within a kilometer of a wolf before you see any behavioral shift or change in their movement." That happens only once every nine days for many of the elk, and for some it's only about once a month, not enough to change their daily behavior. Some areas in Yellowstone might have benefited from the loss of elk, but these changes could not be placed squarely on the return of wolves. Elk declined after the wolves arrived, but human hunting had increased at the same time, and a severe drought occurred throughout the region.

We've been here before. I had trusted colleagues on both sides of this debate, so I called up Os Schmitz, a professor at Yale who has studied the landscape of fear—or, more technically, "behaviorally mediated trophic cascades"—in controlled field experiments for years. What did he think of the Yellowstone studies? He had a nuanced view: "When Yellowstone introduced the wolves, the elk hadn't been exposed to this kind of predator in so long that they couldn't figure out the risk. They were hypervigilant." In other words, when the wolves first arrived, the elk were on the lookout for the new predators in town, which reduced their browsing time. These changes had a positive impact on tree cover and perhaps on the beavers and other species hanging around the river's edge. But the elk soon figured out how to cope. Since wolves are most active at dawn and dusk, they learned to avoid browsing in risky areas during peak wolf hours. But a wolf at rest is not a worry, so elk can feed relatively safely during the day. After they adapted, it wasn't elk behavior but death itself that changed the landscape.

The rain started to fall harder along the Lamar, and McIntyre and I decided to call it an afternoon. As I drove out of the park, I passed a big buffalo lawn that was new to the valley.

MANY PLANT-EATERS, FROM barnacle geese to mule deer to moose, time their movements to follow new plant growth each year. This green wave—the progression of spring green-up from low elevations to high ones—dictates the pace of migrations around the world. Young vegetation provides the best forage, so herbivores follow the wave.

The bison of Greater Yellowstone had recently followed this wave onto the grazing lawns of the Lamar. It's easy to think of them as eating machines, but bison also influence the grasslands when they poop and when they die. "Poop is the main effect," bison biologist Lauren McGarvey told me, "along with the process of grazing itself. We're looking at plants that evolved with grazing for hundreds of thousands of years." Unlike trees, which grow at the top, grasses and sedges grow up from the bottom. Grazing stimulates plants to keep growing.

Of course, that's not the only symbiosis at play. People often think of consumption as simply the removal of biomass or nutrients from the environment, but it's much more complex, as we've seen for species from gulls to whales. Bison glean nutrients from grazing, then deposit them back onto that grassland when they poop. Some of these grasses are consumed just before the plants hit senescence, so grazing keeps them young and green. And these bison-delivered nutrients are in a more accessible form than they would be if they were simply deposited by dead or decaying plants. The nitrogen from the bison poop can be cycled quickly through the soils and taken up more effectively by plants and, eventually, other animals. "So bison actually stimulate plant production," McGarvey told me, "accelerating the rate at which nitrogen cycles through the system."

In addition to his work on wolves, Kauffman has been mapping animal movement. He thinks of migration as something like steps in a seasonal dance. "The mule deer choreograph their movements with the spring," he said. They follow the emergence of new grasses, often to higher altitudes, as the spring progresses, a pattern that's known as surfing the green wave. But when Kauffman and colleagues watched the bison, they noticed that they weren't surfing the wave like other grazers; they were staying in one place for a longer time, even as the green wave kept moving up the mountains. *Wow, they're really not very good at this,* Kauffman thought at first. But then his colleague Chris Geremia suggested that large herds of bison could engineer the grasslands: by eating the taller blades, they kept the grasses in a young, nutritious state.

"When the music stops," Kauffman said, "they can make their own." It was an extended version of the green wave. By recycling nutrients through their poop and pee, bison keep their grazing lawns stable and green, the soils rich in nitrogen and phosphorus. The grasslands and the bison are less vulnerable to seasonal changes.

In early May, bison start to get excited. "Even though their body condition is in the poorest shape of the whole year," Rick Wallen, who was the senior bison biologist in Yellowstone until he retired in 2018, told me, "their energy level is high. The day length is longer. And I think that seeing the new little red calves just gets all of the bison excited. There's a lot more running around in circles and chasing each other. They all want to come over and check the calves out." As the season progresses, this sense of renewal shifts to a focus on breeding. Giant groups of up to eight hundred bison form in Yellowstone as the males chase females. You can hear roars, and there's a lot of pawing at the ground and huge clouds of dust. When the males turn their attention to the females in estrus, Wallen noted, "it's the kind of male-female interactions you recognize in the local bars around the country." But without the restrooms.

"It's not just the bison poop," McGarvey added, "it's also the urine. You could go out this summer and see a very green spot, just a small circle, and tell that was a urine hit from the previous summer—a little nitrogen pulse right on that spot." Sometimes it's more than just taking care of business. Bison mating displays can be spectacular: "The bull thrusts his muzzle into the stream of [the cow's] urine," ecologist Dale Lott wrote in *American Bison*, "then elevates his head, upper lip curled, tongue fluttering inside his mouth, his whole demeanor suggesting a gourmet's appreciation of fine wine. If he goes from lip-curling to tending, the chances are good that the cow will breed sometime that day."

A fine wine indeed, one with effects that linger in the next generation of bison and the bright green grazing lawns of Yellowstone.

OUTSIDE OF YELLOWSTONE and a few remote areas around the West and Midwest, you could travel for thousands of miles and never see a bison. That wasn't the case when European settlers first crossed the Great Plains. In 1871, Colonel Richard Dodge of the U.S. Army noted that he drove twenty-five miles through one immense herd migrating north: "The whole country appeared one great mass of buffalo." The naturalist William Hornaday wrote that bison "were so numerous they frequently stopped boats in the rivers, threatened to overwhelm travelers on the plains, and in later years derailed locomotives and cars, until railway engineers learned by experience the wisdom of stopping their trains whenever there were buffaloes crossing the track."

If you look at the historic range of bison in North America, it's hard to find a place where they didn't roam. The wood buffalo extended into northern Saskatchewan and British Columbia. The southern herd ranged into Durango in Mexico, more than three hundred miles south of the Rio Grande. They lived along the coastal marshes of North Carolina and west to the edge of the Cascades in Oregon. It was a continent of bison, tens of millions strong, when the first Europeans arrived.

That soon changed. More than thirty million bison were killed on the Great Plains in the nineteenth century. By 1875, the great southern herd, which once numbered in the millions, had disappeared. "A few small bands of stragglers maintained a precarious existence for a few years longer on the headwaters of the Republican River and in southwestern Nebraska," Hornaday wrote in *Extermination of the American Bison*. A few dozen held on in the Texas Panhandle. Some of these were captured alive, but even this "miserable remnant" eventually disappeared. After spending a year in Kansas, Dodge wrote: "Where there were myriads of buffalo the year before, there were now myriads of carcasses. The air was foul with a sickening stench, and the vast plain, which only a short twelve months before teemed with animal life, was a dead, solitary, putrid desert."

"Their slaughter has been criminally large and useless," William F. Cody wrote. He had earned the nickname "Buffalo Bill" fifteen years earlier when, as a contract supplier of meat to the railroad companies that were laying down track, he killed four thousand buffalo. After the bison were killed, Michael Punke noted in his book *Last Stand*, they were harvested twice: "Once by the hide hunters; once by the bone pickers." Piled in mounds the size of churches, the carcasses were sold for fertilizer and charred into bone black, the darkest pigment at the time. By 1894, even the bones were gone. There were no herds left to replace them.

What had been lost? The choreography of the bison, of course, the wallows that they formed, the relationships with wolves and other predators, and the long-distance migrations. With the bison gone, the prairie started to disappear too. It had lost the bovid ecosystem engineers and the fires that Indigenous peoples had been setting for thousands of years, fires that opened up forests, promoting prairies, edges, and fire-tolerant oaks, chestnuts, and hickories—nut trees. After the bison disappeared and the prairies shrank, a world without fires and grazers led to a false belief in stability. For much of the twentieth century, people thought that ecosystems inevitably resulted in climax communities, deep-rooted plants and trees that produced old-growth systems with little change.

This desire for stability has made it tougher for bison to return to areas outside national parks. Bison numbers in Yellowstone have increased from a free-ranging herd of fewer than twenty-five in the 1870s to about three thousand in the 1980s and five thousand when I visited. At times, they've been protected; at other times, they've been culled. Their numbers in the Greater Yellowstone ecosystem could be even higher if managers didn't restrict migration out of the national park to prevent the spread of brucellosis, a bacterial disease introduced by cattle and now carried by bison. About nine hundred bison were slaughtered or captured when they left the park in 2021. There's a habitat limit too.

"The areas that bison love, the valley bottoms," McGarvey told me, "are also the areas that we love for housing divisions and ranches." We'll have to look elsewhere if we're going to restore their historical role on the prairies.

"WE STILL HAVE all the songs, stories, and ceremonies, but when you look out there, there's no buffalo to be seen," Leroy Little Bear, a Niitsitapi, or Blackfoot, scholar at Alberta's University of Lethbridge, told me one afternoon from his home in Saskatchewan. Although buffalo (the word many Indigenous groups use for *Bison bison*) played a central role in the lives of his people, there was a problem. Buffalo occupied only 1 percent of their historic range. It was like being a Christian, Little Bear said, "if all the corner churches disappeared."

The people who lived on the Great Plains for millennia had woven these enormous, abundant animals into their lives—sleeping on buffalo-skin robes, wrapping their tepees in buffalo hides, and hoeing with their bones. Fires on the treeless plains were made with dried buffalo dung. "Plains Indians were born on a buffalo robe and wrapped in a buffalo robe when they died," Punke wrote. Buffalo were the foundation of their economy and their culture, much as salmon were for Indigenous peoples of the Pacific Northwest. The Indians used buffalo tails to swat flies, as the tails' original owners had.

"Obviously, the buffalo is a great animal for sustenance," Little Bear told me. "We refer to it as a relative. It is also our teacher. The elders summarize it in our songs, our stories, and our ceremonies. If you were to watch buffaloes move around, you see that it was just like in our ceremonies. They always go clockwise, and it's kind of like they make a big circle of the mountains in the spring, go out further on the plains, and come back around by wintertime, where they can find shelter close to the mountains."

In 2009, a group of elders, including Little Bear and Paulette Fox, who is also known as Natowaawawahkaki (Holy Walking Woman), founded the Iinnii Initiative among the nations of the Blackfoot Confederacy to restore the native herbivores and reconnect young people with their culture. This early initiative grew into the Buffalo Treaty, with more than forty signatories, including the Blackfeet, the Assiniboine and Sioux, and the Blood Tribe. The goal of this first cross-border Indigenous treaty is to restore buffalo across more than six million acres of land in the West.

"The buffalo is a keystone species, culturally and ecologically," said Little Bear. Small mammals and birds use the thick brown bison wool for nests. Insects breed in the wallows, the depressions created by buffalo as they roll in the dust. This grooming helps reduce pests, spread seeds carried in the buffalo coats, and create new habitats for insects and birds. Buffalo cultivate their territory. "You see them running around, and before you know it, it's almost like they plowed the fields with their hoofs so that there would be fresh growth. They're what we refer to as eco-engineers—they bring plants to life that are used for medicinal purposes and food purposes," Little Bear said. Recent studies of bison wallows showed several Native American crops growing tall in the areas opened by the gigantic herbivores. As Indigenous groups followed bison trails, they came upon squash, sumpweed, barley, and sunflowers, which they domesticated in the early days of agriculture several thousand years ago. "When the buffalo returned, it's almost as though you could see a synchronization of plants and ripening for harvest," Little Bear continued. As buffalo moved, people were right behind them. These traditional teachings anticipated recent scientific findings in the choreography of the green wave and the emergence of ancient agricultural plants in wallows.

When I talked to scientists, I often heard the same stories Little Bear had told me but from slightly different angles. Large plant-eaters like buffalo matter—they travel more widely and in bigger herds than many smaller species. They graze more grass, browse more trees, and poop more poop than any other native herbivore in the region. Bison create diverse landscapes, crushing one area and eating everything down to the ground but leaving another prairie patch untouched. After three decades of being grazed by bison, the Konza Prairie in Kansas was transformed. The number of native plants doubled. The bison-grazed grasslands had fewer invasive species and were more resilient to drought. By contrast, cattle tend to be more methodical in their grazing and don't have as big an impact on plant diversity. And unlike cattle, which are raised for consumption, when wild buffalo die, they leave big carcasses, attracting scavengers and passing nutrients back to the prairie.

The reintroduced bison of Konza are fenced in, but new translocations to tribal lands are experimenting with more open borders. Tribes across the West have been given thousands of buffalo, including animals from the herds I saw at Yellowstone. "I think they've sent bison to nineteen tribes," McGarvey told me. In 2020, three bison bulls were FedExed to a tribe on Kodiak Island in Alaska. The goal was to increase genetic diversity in a herd of seventy that roamed free on a nearby uninhabited island. "It helped me realize how important bison are to the tribal partners," McGarvey said. "They were willing to raise about thirty grand to ship three bison to their tribal lands."

Several Indigenous nations have started bringing in buffalo as cultural herds, animals that are not raised for commercial sale. "It's just happening all over the place," Little Bear said, "almost in an exponential way." In the United States, there are more than twenty thousand buffalo on tribal lands in sixty-five herds. He celebrated the return.

Many of the studies of bison have occurred in protected, relatively isolated areas, such as Konza and Yellowstone. But ecological recovery, according to a group of Indigenous peoples, conservationists, and policymakers, will require large herds moving over extensive and diverse landscapes, including long-grass prairies, semideserts, and savannas. "When you start thinking about it, it's all about human survival," Little Bear said. "We owe our existence to the buffalo."

When I asked if there was anything I could do, he said, "Tell people to eat bison burgers." It seemed like a small thing, but perhaps it was a first step in restoring ancient pathways and choreographing a new green wave.

When I went to bed in Cooke City outside of Yellowstone on that Sunday night in 2022, the world seemed normal, if slightly waterlogged. I was looking forward to a leisurely morning the next day before heading out to spend the afternoon in the Lamar Valley observing bison, wolves, and brown bears.

When I awoke Monday morning, reporters were announcing that the rainfall over the past twenty-four hours had been the highest on record for the area. It had melted the five inches of snow that had fallen on the mountains over the weekend, and all that water was now rushing downhill. Mud and rocks tumbled down with it. The rivers jumped their banks in search of new paths, scouring and inundating long stretches of road that had been lined with wolf watchers just a few hours earlier. The road that I took into Yellowstone with McIntyre had washed away. Sewage lines broke. The Yellowstone River crested at 13.9 feet, more than two feet higher than the previous record, set in 1918.

The park was evacuated, vacations were cut short, fieldwork was delayed. The northern entrances were closed. Nine bridges were torn from their stanchions, isolating McGarvey and other biologists in the community of Gardiner. Several Yellowstone employees lost their home after it fell into the Yellowstone River and floated downstream. The park superintendent, Cam Sholly, called it a thousand-year event, adding, "Whatever that means."

River systems are dynamic, and flooding isn't all bad. It can restore isolated habitats, connecting them to the main stem of a river, allowing fish, amphibians, and insects to move. But there were immediate concerns that climate change might have played a role in the June storm, providing a glimpse of the oncoming chaos: bigger fires, hotter temperatures, longer droughts, and larger floods. It's hard to pin the damages from any event solely on climate change, but scientists expect less snowpack, increased annual precipitation, and drier summers, which can elevate fire risk in the Greater Yellowstone.

Rescue boats were being trailered in from the east, the only road out of town. The hotelkeeper offered to give me all my money back if I left town. I had planned to stay for a week, but when a seasoned local, especially one who happens to be a first responder, tells you it's time to go, I suppose it's time to leave.

Andskotinn, I heard my inner Icelander swear.

We jostled over a dirt road in the Maasai Mara in a Land Rover. Except for the occasional muddy puddle, the savanna was dry as a bone. We stopped along a stretch of the Mara River. There were some steep banks carved into the sides.

Amanda Subalusky and her husband, Chris Dutton, have been working in Kenya for fifteen years. They started out as technicians examining water quality in the Mara River. In early studies, they noticed that the stretches of river that ran through the conserved areas of the national park had higher levels of nitrogen and higher bacterial counts than the towns and farmlands did. At first, this didn't make sense. (Far more discoveries, a colleague once noted, start with a "Huh, that's weird," than with a light-bulb moment.) The pattern was the opposite of what we see in North America, where farms and lawns are often the biggest sources of nitrogen and phosphorus runoff. Shouldn't wild areas be clearer, with lower levels of nutrients and fecal bacteria?

And then it occurred to them (I like to think of it as a brown flash): Hippos are like lawn mowers, slowly moving their enormous heads as they feed on the savanna, creating big hippo lawns with short, well-trimmed grass. They are well known for spraying feces, with their tails swirling around like helicopter blades on land. "They'll kind of poop in each other's faces," Subalusky said as she pointed to some feces along the trail. They move about a mile or so every day from the savanna to the rivers to rest and cool off in hippo pools. "There's *so* much hippo dung," she continued, "it just blankets the bottom of the river. The rocks are covered with it." It was a hippopotamus conveyor belt.

The nutrients in the dung spike growth in algae and plants in the river. It was unclear if it was the algal bloom from the nutrients or the poop itself that attracted the fish and invertebrates, but Subalusky and Dutton had at least one hint. "We caught this two-and-a-half-foot-long catfish, a big river monster," Subalusky said from the passenger seat; she had on a light pink shirt, and sunglasses were perched on her head. "When we looked in its stomach, it was all just hippo shit."

A couple of hippos lounged at the surface, sunning themselves. One yawned in the middle of the river. A Nile crocodile swam downstream. And then there was a glitch. The animals disappeared.

"I'm sorry, Joe," Subalusky said. "Let me see if I can turn you around." Dutton, with a thick wiry beard, loomed in the driver's seat. Subalusky held me at arm's length just next to the side-view mirror.

Our connection cut out for a moment. I had been traveling with them via FaceTime. Soon after it became clear that my plans to accompany the Yellowstone bison biologists into the backcountry would be washed out by the floods, I contacted Subalusky and Dutton. Concerns about carbon, COVID-19, canceled flights, family, and travel weariness after three months of being on the go kept me at home. I was in the Maasai Mara, and I was in my office; the dog was barking to go out for a walk in the kitchen upstairs.

Subalusky and Dutton later realized that the hippos we were watching supplied more than just nutrients to the river. "The hippo dung and the gastrointestinal microbes shaped the river into a good place for gut microbes," Subalusky said. The microbial communities in the waters outside of the hippo pools came from different sources and had lots of diversity, but the communities at the bottom of the hippo pools, black with dung, were more like those inside the hippos themselves. The hippos defecated in the pools and drank the water, forming a vast microbial community that Dutton and Subalusky called a *metagut*. The hippo's microbiome was still functional even after some of it had been released, breaking down grasses and continuing biogeochemical processes. After the pool flushed out, the microbial community would resemble the rest of the river's for a while. Then the hippos would return, and the process would start all over again.

The hippos were having a regular influence, like whales and seabirds, through their poop. But there was also an unexpected pulse of nutrients about once a year. "We showed up at the river one day, and there were all these wildebeest carcasses," Subalusky said, "just hanging out in the eddy of the river." Subalusky and Dutton had known about the wildebeest migration, among the largest in the world, with about a million wildebeests traveling across the Serengeti and Maasai Mara each year. But they hadn't expected a mass mortality. The carcasses piled up. Vultures flew in. Hyenas fed at night.

Seventy-five years ago, such crossings were rare. Wildebeest populations almost disappeared in the mid-twentieth century after they were exposed to rinderpest, a virus then common in livestock. The treatment of the disease and its eradication in 2011 led to an exponential increase in the number of wildebeests, also known as gnus. They're broad-horned and barrel-chested, with faces like Abraham Lincoln—or so Laurie Anderson sings. The survivors, along with hundreds of thousands of gazelles and zebras, migrate across the

African plains every year. Their grazing helps rebuild the soil carbon in the Serengeti; ungrazed grasses can build up fuel loads that make vast areas vulnerable to fire. The return of the wildebeests resulted in fewer fires. Insects helped move carbon from their dung into the soil. The Serengeti ecosystem switched from being a major carbon source, released through fires, into a carbon sink, absorbing several million tons of carbon dioxide annually.

For much of the year, there are few wildebeests in this area, but that all changes in June. "It's like you wake up one morning and the whole thing has been sprinkled with black pepper," Subalusky said. "You can't really wrap your head around it. You'll come to some vistas where there are wildebeests literally as far as you can see."

The wildebeests cross the Mara River multiple times a year. When the river is low, they might just get their ankles wet. But at times of high discharge, the crossing is perilous. Individuals balk at the bank, eyeing the white water, but once one of them leaps, a herd mentality can take over. Wildebeests are not great swimmers. Many never make it to the other side.

"The most dangerous time to take water samples is just before the wildebeests show up," Subalusky said. "Nile crocodiles know where the wildebeests cross, and you'll see them gathering—thirty or forty large crocodiles congregating around the crossing sites."

Subalusky and colleagues tested the water, caught fish, and collected microbes. "Then all of a sudden," she recalled, "we're sitting on the bank of the river and seeing hundreds of wildebeest carcasses in the river."

So there were predators and scavengers showing up at the crossings. And it stank. The smell of death is largely the chemical ethanethiol. Humans are so sensitive to this sulfur compound that we can detect it at one molecule in a billion. (It's added to propane to alert you that the gas is on and keep you from striking a match. The story goes that oil workers in California noticed that turkey vultures would gather at gas leaks because they detected this smell, so the compound was later mixed with odorless fuels. We add the smell of death to avoid it.) The scent of rotting flesh might make humans wrinkle their noses, but the smell appeals to vultures, flies, and other scavengers. Ethanethiol and other volatile odorants are so attractive that some flowers, such as dead horse arum, use them to entice blowflies into pollinating them.

What is the impact of hundreds of wildebeest carcasses piling up in the river? Subalusky and Dutton used camera traps to count scavengers. Vultures flew in from miles away; hyenas edged into the river. And then there were the flies, which attracted ibis and mongooses. "I've got a photo of a juvenile crocodile basking on these wildebeest carcasses," she told the *Atlantic,* "and I think he looks so happy."

Subalusky estimated that at any given time, the standing stock of bones in the Mara was equivalent to that of about fifty blue whales. As the skeletons decayed, phosphorus leached into the river, and catfish fed on the algal mats that covered the bones. Does this ring a bell? You guessed it. "I was really familiar with the salmon migration," Subalusky said. Bob Naiman became a mentor of hers when she was working on her master's, so she had read a lot about salmon.

Subalusky's work on animal subsidies in Africa beautifully illustrates the flow of nutrients, like a circulatory system, moving in pulses around the planet. Watching wildebeests and reading over early accounts from North America got her thinking about another long-term migrant. Mass drownings of bison were common on the Great Plains in the spring, especially when the trails crossed over rivers. In 1795, the fur trader John Macdonell spent the day counting carcasses of bison that had drowned in the Assiniboine River in Manitoba. He found a total of 7,360 before he crossed the river on the backs of the carcasses and stopped for the night. Subalusky recognized a kindred spirit. She and her colleagues estimated that about two hundred thousand bison died in the rivers of the Great Plains each year back when there were tens of millions migrating across rivers. This would be about a hundred thousand tons of bison annually; depending on your preferred metric, that's about a thousand blue whales or four Statues of Liberty. The bones they left behind were long-term reservoirs of phosphorus just like the wildebeest bones in the Mara, leaching out slowly and being replaced each year.

And what about movements around the globe? Martin Wikelski, the director of the Department of Migration of the Max Planck Institute, gave a seminar at Yale when Subalusky was a grad student there. "It lit a fire under me that is still burning," Subalusky said. Wikelski's website Movebank lets you visualize data from more than two billion locations from thousands of tags placed on all kinds of animals, everything from sea turtles to blue whales to free-ranging dogs. Their movements river the globe in neon lines from the Arctic to Antarctica and everywhere in between. Each dot, each line, is a particular animal, like a cell in the bloodstream—a wolf moving through Denali; a seal swimming through Scottish waters. You can follow the path Princess, a white stork, took as she traveled from Germany to South Africa and back every year from the time she was a preadolescent bird in 1994 until she died in 2006.

"It just sends goose bumps down me every time I watch it," Subalusky gushed. She has considered how this movement across the planet delivers animal subsidies. There's a donor ecosystem, often highly productive and nutrient-rich—for seabirds it might be the ocean, for whales high latitudes, and for hippos grasslands. Many animals move in and out of these areas, perhaps because there are too many predators in their usual homes or because they need a place to rest, mate, nest, or nurse. The recipient ecosystem tends to be less nutrient-rich; it might be an island, a river, a tropical beach. The animals poop, pee, and maybe give birth or even die in this new habitat, and in doing so, they enrich the ecosystem.

"Billions of animals are adapted to a traveling life," Swedish biologists Thomas Alerstam and Johan Bäckman wrote in *Current Biology*, "making regular return migrations between more or less distant living stations on Earth by swimming, flying, running or walking." The wildebeests and hippos of the Maasai Mara were just a small part of this circulatory system. And much as poor circulation can lead to the loss of a limb, the decline in animal populations and migration can cut off the nutrient supply to some of the world's richest habitats, making them less productive and more vulnerable to climate change.

THE STORMS THAT hit Yellowstone created the highest floods on record, and it seemed likely that climate change and global weirding—an increase in weather-related extremes, from flooding to ice storms to hurricanes—were contributing factors. As the water rose, I retreated to Cody, home of Buffalo Bill. There was water up to the edge of the Clarks Fork bridge. A logjam pressed up ominously against the retaining wall.

The flood would be bad for those residents who lost property or their summer income. But it might be good for the wildlife that wasn't directly harmed by the storm. As I drove on, I imagined the bison, wolves, and grizzlies wondering what the hell had happened to all the cars that had snaked through the Lamar Valley the day before. Maybe now they were becoming less stressed, like right whales after September 11, 2001, when marine traffic stopped and the oceans went quiet for a while. McGarvey wondered if the wolves and grizzlies were exploring new areas, relying on human roads, as they might have during the COVID lockdowns.

Seen from a distance, as I traveled southeast on the Chief Joseph Highway out of Yellowstone, the large black cattle resting on the hills looked like bison, dark against the green-sculpted, almost barren hills. But behind the barbed wire that fenced them in, did they perform the same ecological role that their wild migratory cousins once played? "Cattle certainly consume grass," Rick Wallen told me from his home in Bozeman Pass, "but cattle are not prey for predators because ranchers shoot those predators." Cattle don't use the landscape in the same way that bison in large numbers do. They tend to be thirstier, spending more time by the rivers and streams, enhancing erosion and runoff. Humanity has devoured large areas of the western United States, to the detriment of bison conservation and restoration.

"A herd of bison crossing an interstate highway in the way that Colonel Dodge described bison crossing the railroad in the late eighteen hundreds"—a herd twenty miles long and two miles wide—"would create a completely different social impact today." Wallen doesn't think society would tolerate wild bison outside of places like Yellowstone, tribal lands, and the new American Prairie reserves along the Missouri River. Battles continue over whether to reintroduce bison to the million-acre Charles M. Russell National Wildlife Refuge in northern Montana, a plan that has the support of Indigenous groups but is opposed by some ranchers. "It's a hard row to hoe," Wallen noted. "I'm pessimistic about the possibility of our society coming together, as polarized as we are, to do something significant like we've done at Yellowstone, but I haven't lost hope."

The restoration of wild animals could make them as newsworthy as the weather and as dependable as the seasons, but what happens when even the climate is undependable? As important as animals are in a place like Yellowstone, climate chaos—torrential rains, earlier snowmelt, extended droughts—could overwhelm their influence. There has been a thrilling rise of bison, wolves, beavers, and other native species in the nation's oldest park in recent years, which promises to help restore ecosystems and landscapes, but extreme events challenge the prevailing norms of wildlife conservation. In 2070, Yellowstone is expected to be five or six degrees warmer than it was in 2000, and it will be up to ten degrees higher by the end of the century. Some of Yellowstone's animals will adapt, others will move on, and still others might disappear entirely.

Can animals ever be as powerful as they were before humans dominated the planet? It's worth considering how we got here.

5

Chicken Planet

When I walked out of my office the other afternoon, car horns outnumbered birdsongs. I heard the tap of hammers rather than woodpeckers. The only migration I saw was done by commuters on their bikes or in their trucks or cars.

You might think that most large animals have disappeared from Earth, but in sheer weight, there are more mammals on the planet now than there were before humans started hogging the sunlight. The issue? Most of them are the four-legged animals we love to consume. Cows, pigs, sheep, and other animals we eat account for 60 percent of all the mammals on Earth by weight, about a hundred million tons. We humans make up 36 percent, or sixty million tons. And the total biomass of wild mammals? Just seven million tons, or 4 percent.

When I first read this in the *Proceedings of the National Academy of Sciences,* I rubbed my eyes. Was it really possible that all the rhinos, hippos, elephants, moose, sea otters, bears, and even great whales were outweighed twenty-five to one by humans and our livestock?

Global biomass of mammals: Humans make up 36 percent of all mammals by weight, domestic animals 60 percent, and wild mammals, including great whales, just 4 percent.

Yep. Our species has hunted out about 85 percent of all wild mammals since we took over the African savannas and migrated around the world. As the large animals disappeared—which happened about ten thousand years ago—humans replaced them with domestics. No wonder wild animals are dismissed as ecosystem engineers. In the Anthropocene—the current era, which began when humans started having an overwhelming environmental impact on the planet—wild mammals are little more than a rounding error.

It's the same for birds. Chickens, turkeys, and ducks outweigh their wild relatives—including all the ostriches, condors, emperor penguins, fulmars, bald eagles, barred owls, herring gulls, bluebirds, chickadees, and grasshopper sparrows—three to one. The slice left for wild birds isn't as thin as it is for mammals, but only about 30 percent of avian biomass is wild.

There are ecological consequences to these changes. Many animals are raised on intensive farms rather than on pastures or grassland. Liquid manure from these factory farms releases ammonia into the atmosphere, a cause of acid rain and nitrogen pollution, and it can pollute groundwater with nitrates. Cows ferment cellulose and grain in the fields, feedlots, and dairy barns of the world; as a by-product of this chemical reaction, each cow releases more than two hundred pounds of methane into the atmosphere annually. It adds up. Beef and milk production accounts for more than a third of human-caused emissions of methane, which is about eighty times more powerful than carbon dioxide as a greenhouse gas.

We've altered the world so much that we've broken a basic law of the ocean. Across the size spectrum, from microbes to whales, the biomass of each class of marine life has historically weighed the same. That is, if you were to calculate the collective weight of all the small zooplankton on the planet, each of which typically weighs less than an ounce, it would equal the mass of all the organisms in the next size class up—say, forage fish—and so forth, up through the megafauna like sharks and whales. This pattern, that the abundance of an organism is inversely related to its body size, is called the Sheldon spectrum. Put another way, krill are ten thousand times smaller than cod, but they're also ten thousand times more abundant than cod. Industrial fishing and whaling changed all that. The bigger animals have been largely extirpated by humans, and the smaller ones dominate the oceans. Whales, other marine mammals, and the largest fish have been reduced by almost 90 percent since 1800. Never mind the Anthropocene—we've entered what marine ecologist Daniel Pauly calls the Myxocene, "the Age of Slime" (*muxa* means "slime"). An ocean devoid of fish, sharks, seabirds, whales, and large invertebrates is a world without marine voyagers, an ocean that provides neither awe nor edible creatures.

Despite these changes, humans have not set up an "independent sovereign kingdom," as environmental historian Donald Worster noted. "In truth, there is no escaping the ecological matrix." Before we fenced out the wild and killed off more than 80 percent of the mammals, fish, and birds, there was a balance between the nitrogen and phosphorus that leached downstream and that which flew, swam, ran, or crawled around the globe via animals, distributing these nutrients in the form of feces, meat, and bones.

Once we lost this balance, we had to look elsewhere for these nutrients.

IN 1802, THE Prussian explorer Alexander von Humboldt arrived on the Pacific coast of Peru after years of travel across the Americas. He had recently climbed the volcano of Chimborazo, then believed to be the highest mountain in the world. Though unprepared for the jagged climb, he and his companions had reached the highest altitude ever recorded by humans: 19,413 feet. As he stood on Chimborazo, Andrea Wulf wrote, Humboldt looked down at the changing types of vegetation he had encountered at each level and saw plants in a new way, viewing "nature as a global force with corresponding climate zones across continents."

Back down on Earth, Humboldt approached a seabird island that was valued by the Inca as a source of fertilizer. The seabirds on the island defecated just before takeoff, and the locals had observed the buildup of excrement, eggshells, and carcasses over decades and centuries. They called the chalky substance *wanu,* the Quechua word for "bird shit." It was so valuable that under Inca law, disturbing seabirds during their nesting season or killing them at any time of year was punishable by death.

Humboldt may have revolutionized the way we see the natural world, but he was a poor ornithologist, according to Gregory Cushman, author of *Guano and the Opening of the Pacific World.* Humboldt looked at the dry substrate the locals had collected and questioned their interpretation; he considered it the result of some ancient catastrophe, like the coal beds of Europe. Although the smell of ammonia along the docks was so strong that he erupted in fits of sneezing, Humboldt was intrigued enough to bring some South American guano back to France. He turned it over to a close friend, the chemist Louis Vauquelin, who found it had surprisingly high concentrations of nitrogen-rich uric acid and recognized its potential to boost soil fertility. Later experiments showed that guano was a far better fertilizer than pig or cow manure, but Humboldt and his colleagues still weren't sure of its source.

Europe was desperate for these nutrients and had been for centuries. In medieval England, peasants could graze their sheep on the land of nobility, but they faced severe punishment if they were caught removing the droppings. When peasants moved to the city, their own urine and feces were unleashed in urban areas. The toll of infectious diseases became apparent, leading to the sanitation revolution, and sewage was moved out of the cities and released into lakes, rivers, and, eventually, oceans. The valuable nitrogen and phosphorus were shunted off with the sewage. Animal carcasses, once left in rural fields, were transported to urban butchers and larders. Rag-and-bone men roamed the city streets. The bones they gathered were carved, rendered into glue, ground into fertilizer, or used to refine sugar; the rags were mostly used to make paper.

In the nineteenth century, English tomb raiders and bone pickers traveled long distances and unearthed thousands of skeletons of soldiers and the horses they had ridden into the battles of Leipzig, Waterloo, Austerlitz, and elsewhere. England became the world's greatest trafficker in human bones, its citizens carrying away generations of skeletons from the catacombs of Sicily and mummies from Egyptian tombs. Many were shipped to the port of Hull and ground in the bone mills of Yorkshire.

There was a limit to the number of human bones that could be unearthed in Europe, but the guano supplies—at least at first—seemed endless. The Inca of South America had had it right all along—guano was a living resource. Millions of Peruvian cormorants, pelicans, and boobies fed in the nutrient-rich waters of the Pacific. Their hatchlings formed crater-shaped rings of guano that the Inca called *quillairaca*—"the moon's vagina." The seabird islands had a hot, almost rainless climate, so much of the poop stayed onshore rather than getting washed out to sea, forming a natural mineral deposit.

Over the years, the splatters of bird poop gained mass, and they started to resemble the rocks themselves. On some Peruvian islands, the layers of guano were two hundred feet deep—eighteen stories of seabird dung. The highest concentrations of nitrogen and phosphorus ever measured on Earth's surface were found beneath these seabird colonies.

CUSHMAN SETS THE start of the Anthropocene at 1830, with the first shipment of nitrate fertilizer, soon followed by guano from South America to England. Guano was pickaxed out of the ground, loaded onto carts, and lowered to waiting vessels. As the international guano trade grew, tens of thousands of laborers were brought over from China to work beside local convicts and debtors, many in conditions of slavery. American, British, and German ships soon crowded the Peruvian coastline, eager to fill their holds with the cargo.

Guano was as essential to the Industrial Revolution as fossil fuel, Cushman argues, the European discovery marking a pivotal moment in human history. As they uncovered its value, Humboldt's colleagues considered the export of guano akin to a moral mission; the spread of nitrogen and phosphorus was good for all people. Soil enriched with guano kick-started input-intensive farming, increasing the productivity of crop- and pasturelands, allowing urban populations to grow and thrive. A common dictum in the nineteenth century was "Make two blades of grass grow where one grew before." Guano helped make that possible. The fertilizer reached Europe at the end of one long era of human history, a time of stability, small towns, and subsistence farming, and helped galvanize another—an era soon dominated by machine production, the steam engine, and agriculture produced for urban markets.

The nineteenth century was the age of guano, and the search for this powerful fertilizer drove global commerce, starting on the ancient coasts of Peru and soon encompassing some of the remotest islands in the world. The significance of this trade is almost impossible to overstate. In 1841, Britain imported about two tons of Peruvian guano. Four years later, it imported nearly 220,000 tons. Moving nutrients continues to dominate food production today. Fertilizer from guano altered the nitrogen cycle in the Northern Hemisphere and caused a grasslands revolution in the outposts of Australia and New Zealand. Southern pastures became more fertile, and the lamb and beef raised on bird dung were exported to Europe, the United States, and the Middle East. "Peruvian guano mainly served northern consumers of meat and sugar," Cushman wrote; it wasn't fighting world hunger.

FOR A WHILE, Peru's seabirds were the most valuable ones on the planet—billion-dollar birds. Peak guano hit around 1870. Eventually, overfishing depleted the anchoveta that Peru's seabirds relied on, and rats, cats, and pigs devastated many of the birds of the guano islands of the Central Pacific. In the twentieth century, remote islands became popular as nuclear testing sites; thousands of nesting seabirds were killed, some entirely denuded of feathers, others with the eyes burned out of their pointed heads. Millions of birds were repeatedly exposed to atomic tests in the years following World War II.

Even before overfishing and the nuclear age arrived on these distant islands, it was clear there wasn't enough guano; nothing was ever enough. Once again, people turned to animals for fertilizer, but this time the bones of ancient creatures were the source of phosphorus. By the late nineteenth century, Florida's Bone Valley, rich in the fossils of Miocene mammals and marine creatures, became a center of extraction, with numerous mines and phosphate-processing plants, as well as towering chalk-white stacks of gypsum, the waste from the mines. Early mining was done by hand, with pick and shovel. Zora Neale Hurston described the work, mostly done by Black men, in her autobiography: "They go down in the phosphate mines and bring up the wet dust of the bones of pre-historic monsters, to make rich land in far places, so that people can eat. But, all of it is not dust. Huge ribs, twenty feet from belly to backbone....Shark-teeth as wide as the hand of a working man." Laborers spent ten to twelve hours a day knee-deep in water as dynamite blew rocks from the walls. Armed guards patrolled on horseback to ensure that no one left for a more lucrative job elsewhere—and that convicts who had been leased by the State of Florida to work the mines didn't escape.

In the twenty-first century, there has been rising concern that we have reached peak phosphorus. The United States trails China and Morocco in phosphate mining, and guano is still harvested and sold in Peru and Chile. But the sources of the element, whether ancient or newly deposited, could once again be a bottleneck for agriculture. The debates continue about when or if a global phosphorus crisis will arrive, but seabirds, even in their diminished numbers, continue to fertilize islands, shorelines, and coral reefs. By one account, the movement of phosphorus and nitrogen to coastal ecosystems is valued at $473 million a year, and it's up to $1.1 billion if you include tourism and fisheries. Peruvian cormorants, pelicans, and boobies are not as common as they once were, but they, along with other seabirds, are still billion-dollar birds.

WE LIVE IN the age of the bird. But now they are mostly flightless, and many never see the light of day. About fifty billion chickens are slaughtered annually across the world, and more than eight billion broilers are raised in the United States each year. Across the Broiler Belt—an area stretching from Maryland to Texas, though lots of birds are also raised in California—chickens raised for slaughter often live in flocks of ten thousand or more, spending their entire lives indoors. (Not just broilers, but egg layers too. Laying hens often live in tiny wire cages, stacked side by side or on top of one another.) From the time they hatch until the moment they die, they never experience fresh air or natural light; their entire existence takes place in a barren landscape of poultry, shit, and feed. "The moment you set foot in one of these warehouses, this toxic stench hits you like a ton of bricks—the air is acrid and suffocating," Leah Garcés, president of Mercy for Animals, told me. "You cough, your eyes water, you can't breathe. You can't imagine tolerating this for more than a short time, let alone spending your life here."

These broilers are lifers, nonetheless, and they barely resemble their ancestors. In the six thousand years since chickens were domesticated in Southeast Asia, we've changed their skeletons, genes, and epidemiology. Their bones are three times as wide and twice as long as the red jungle fowl, their undomesticated counterparts. They grow three times as fast as their wild relatives and die much younger, slaughtered at around six weeks. During the medieval period, chickens doubled in size through selective breeding on the farm. In the twentieth century, their size has increased fivefold. Domestic chickens have lower genetic diversity than their wild relatives and a mutation in a hormone receptor that allows them to reproduce year-round. Genetic manipulation and intensive, industrial farming practices have given rise to a world where chicken is plentiful—and shockingly cheap, if you don't consider the true costs behind their production.

Almost half a billion never make it to the high-school lunchroom or between the buns of Chick-fil-A—they're too sick or weak to survive the six or seven weeks until slaughter. Even those that do are plagued by illness and injury. "Tens of thousands of defecating birds, many debilitated by unnaturally fast growth, have no choice but to wallow in their own waste, which builds up flock after flock," Garcés continued. Their feet and chests become raw with bedsore-like burns. "When you buy chicken for consumption at a restaurant or a grocery store, this is what you're eating."

For these chickens, and similarly for the cows, pigs, and other animals raised in industrial farms, life is a wire cage, a metal crate, a super-flock with little room to move, their offspring taken away as soon it's efficient to do so. With such large numbers, they can act as incubators and superspreaders for bird flu and other zoonotic diseases.

We've packed so many landfills with those extra-large chicken bones that drumsticks, wishbones, and wings are likely to be a key marker of this age in the fossil record. These bones can also track ancient human migration. In an archaeological site in Chile, researchers found a single chicken bone, which they dated to AD 1400. The DNA sequence matched that of chickens found in Vanuatu and other South Pacific islands. The bone provides evidence of the movement of people from Polynesia to South America, showing that chickens got to Peru before the Spanish conquistadores in 1532.

You can call ours a chicken planet. You can call it a cow planet. But no matter how you look at it, life has been domesticated. The story of human movement and influence across South America and the planet goes back much further than 1532.

In 2010, when Chris Doughty was doing postdoc work on carbon forest dynamics at the University of Oxford, he came across the skeleton of a giant ground sloth on Fossil Way, the yellow-bricked entry hall in the Museum of Natural History in London. A resident of South America, *Megatherium* was over thirteen feet tall with long claws and a deep, jowly jaw.

Larger than elephants, giant sloths towered over the ancestors of the tapirs and peccaries that remain in South America today. Unlike their modern tree-dwelling relatives, giant sloths weren't slow. These enormous vegetarians moved freely about the savannas, likely depending on foregut fermentation to break down the cellulose in their diet, much like bison, hippos, and other grazers. This type of fermentation generally doesn't work for critters bigger than a hippo, in part because the methane buildup in the gut can get dangerous. Giant sloths were so big—up to thirteen thousand pounds—it's amazing that they didn't explode after a big leafy meal, one biologist joked.

Doughty looked up and wondered: In all their eating, pooping, and dying, how did these giant sloths, which could reach sixteen feet above the forest floor, structure the forests and savannas?

He didn't wonder only about quick-moving sloths. Before humans emerged from Africa, Earth was ruled by giants: enormous tortoises, elephant birds, moas, armored glyptodons—part armadillo, part VW Beetle—and other Seussian and Borgesian creatures. The toxodon looked like a hippo mixed with a rhino, according to Doughty. Hornless, with big mule-like teeth and shaggy fur, they were once the most common hoofed mammal in South America. There were predators throughout the Americas, too, saber-toothed cats, American lions, enormous bears and wolves, and flightless terror birds that

Profile of a few Pleistocene species of the Americas, including the giant sloth (far right), glyptodon (lower right), mammoth (with tusks), camel *Aepycamelus* (center), and horselike *Moropus*. Bison, on the left, were smaller than most of these megaherbivores. They are the only ones in this figure that survived the Pleistocene extinction.

stood nine feet tall and could kill their prey with a single blow. The animals fought for every morsel. In the fossil record, many of these carnivores' teeth were worn down to the bone, probably because they were desperate even for the marrow.

In addition to the elephant-size sloths, there were lots of actual elephants too. Proboscidean diversity peaked about three million years ago, when thirty-three species of elephants and their relatives roamed the world. There were mammoths, mastodons, gompho-theres, and tetralophodons, which were ten feet tall with long tusks and small ears. Abundant and ecologically innovative, members of this large branch of the evolutionary tree thrived until it was cut off about a hundred thousand years or so ago, when climate shifts and human hunting caused a catastrophic decline in global diversity and only a few species survived.

How did the daily movements and seasonal migrations of these giants, pooping and dying for millions of years, affect the landscape? Doughty and colleagues started looking in South America, land of the sloths and toxodons. "There were concentration gradients ev-erywhere," he said. Large mammals and other animals can move long distances across these gradients, from areas of high nutrient concentration to areas with fewer nutrients, much as whales move nutrients from high to low latitudes. "I always think about this as an integrated system—there are certain elements, like phosphorus, that are taken up by plants and needed by animals," Doughty told me from his office at Northern Arizona University, where he is now an associate professor, "and that's the critical junction. First you need to understand how efficiently the elements move from the soils to the leaves, and then from the leaves to animals. Once the animals con-sume the foliage, they'll walk a bit and poop it out."

Doughty quantifies this last movement as Φ, or phi, the diffusion term for dung—the lateral transfer of nutrients by animals from one ecosystem to another. In newer lands, like Surtsey, nitrogen tends to be limiting, since volcanic rock is rich in phosphorus, but in older regions, mountain chains like the Appalachians and tropical forests like the Amazon, the phosphorus gets depleted. "Basically, it's just old soils," he said. After millions of years, much of it runs into the rivers and eventually the ocean.

New phosphorus can be released by natural weathering, through the erosion of rock and minerals by water, ice, and wind, evocatively described as Aeolian dust. One of the only ways—short of a new volcano or favorable winds—to get phosphorus back uphill and upstream is through the dung, flesh, and bones of animals.

The scats of fruit-eaters are often rich with seeds, like chips in a warm dough of nitrogen and phosphorus fertilizer. The carcasses of giant sloths, gomphotheres, and toxodons, frequently weighing several tons, were enormous nutrient packets too, though decomposers and plants had to wait for these long-lived species to die.

"You have very low phosphorus-loss rates in the tropics because everything is evolved to recycle it really well," Doughty continued. You can see this at play in the Amazon. If you poop in the woods after lunch, it will be gone before dinner, broken down by the insects and humidity. In a dry place like the Kalahari, feces could probably hang around for months. In tropical rainforests, the phosphorus that animals move is quickly reincorporated into nearby plants.

Why do we care about phosphorus in places like the tropics, again? "Without phosphorus, growth rates slow," Doughty said. "Plants are less productive. There are fewer fruits and flowers. But once you add phosphorus, photosynthesis increases, growth rates climb." And plants can devote more energy to reproduction.

Animals need phosphorus too, of course, but they take only what they need and release the rest through feces and urine on a regular basis. Large-bodied animals are big, concentrated sources of nutrients with lots of phosphorus in their bones, but they can live a long time, and even when they die, it can take a while for their bodies to decompose. Carcasses might offer a larger nutrient pulse than poop, but Doughty's phi calculations indicated that the movement of phosphorus and other nutrients in dung far outweighed the nutrient footprint of a carcass.

It was Doughty, as far as I can tell, who first saw these nutrient pathways as the world's circulatory system. As we've seen, animals are the beating heart of the planet. Giant sloths, armadillos, and mastodons helped spread phosphorus, nitrogen, and other vital nutrients across the continents, often dispersing them from nutrient hot spots. These days, whales swim them across oceans, seabirds fly them onshore, and insects move them over the fields.

WHEN *HOMO SAPIENS* first diverged from our earlier human ancestors about 300,000 years ago, there were more animal, plant, and fungal species on Earth than in any of the previous 4.5 billion years. Humans are both the youthful products of biodiversity and the prominent cause of its demise.

The number of species on the planet—including birds, amphibians, fish, reptiles, invertebrates, and plants, among others—has been going down ever since humans moved out of Africa. The great extirpation of large plant-eaters in the Americas started in the late Pleistocene: peak extinction, especially among savanna animals like giant sloths, glyptodons, mammoths, and mastodons, occurred about twelve to thirteen thousand years ago, when humans arrived and climate change intensified. One hundred and seventy-eight large mammals, terrestrial megafauna, went extinct at the end of the Pleistocene. As these species disappeared, so did their predators, like large saber-toothed cats. Species that depended on these large animals, living on their skin, perhaps, or in their intestines, also disappeared, as did those that relied on the wallows they made or the grasslands they maintained. Vampire bats that drank the blood of the largest animals, dung beetles that fed on their enormous poop, and vultures that relied on the carcasses soon disappeared too. Ecologists call these losses coextinctions—and they were vast.

Though there is some disagreement over the role of climate change versus overkill in the Pleistocene extinctions, I find the work of Jens Svenning at Aarhus University, Felisa Smith at the University of

Mexico, and many others convincing: the evidence points to over-kill by humans as the primary driver of megafauna extinctions at the end of the Pleistocene. "I think the debate will never go away," Svenning warned, "just like people keep discussing why the dinosaurs went extinct." Doughty described the argument as a "blood feud."

That our ancestors could kill mammoths—many much larger than the present-day elephants on the African savanna—by digging pits and sharpening spears *is* an astonishment. All seven mammoth species went extinct after the arrival of humans, and the loss of large-bodied mammals from ecosystems is a key sign of human migration. The last major showdown between upright apes and proboscideans occurred in South America, according to Doughty. It was the only large continent free of humans up until about twelve thousand years ago. The gomphotheres, a distant relative of the modern elephant, lost the battle and went extinct about eleven thousand years ago.

As Doughty has noted, we traveled the world on the flanks of wild megafauna, from ground sloths to mastodons. We look for the biggest animals, kill them, and then work our way down; the size of land animals shrank in Africa, then across Eurasia, Australia, the Americas, following in the wake of human movements. And once they were gone, we used their habitats for food. Organisms compete for resources—the ecological term is *competitive displacement*—so once the large animals disappeared, humans soon, at least in geological terms, took to replacing them with cows, chickens, sheep. There is only so much solar radiation to support plants and animals, wild or domestic. The extinction of mammoths, ground sloths, glyptodons, and other Pleistocene ecosystem engineers helped open up the modern human world.

In a land without mastodons, the white-tailed deer is king. This pattern of downsizing wild animals would eventually extend to the oceans too. Landlubbers and sea dogs alike live in some variation of the Age of Slime.

We see these population declines in the fossil record but also in the genes of living animals. Svenning and colleagues examined the genomes of more than a hundred surviving large-mammal species. The population history of animals remains encoded in their DNA—we can still see evidence of a population bottleneck in humans that occurred about seventy thousand years ago. In the case of many large mammals, such as bears, mountain lions, bison, and deer, genetic variation indicates that populations fluctuated up and down for about a million years. About a hundred thousand years ago, genetic variation, linked to population size, declined. This is exactly when humans came on the global scene. In addition to the widespread extinctions of many of the largest species, other species on the planet went through a population bottleneck that continues to this day. Before that time—before the arrival of humans—Svenning said, "ecosystems worldwide were filled with animals in a way we can't even imagine." We've now come to accept landscapes without large animals as the natural state of things.

In South America at the end of the Pleistocene, 70 percent of animals that weighed more than twenty pounds (so beagle-size or bigger) went extinct, including the gomphotheres, giant sloths, and glyptodons. Their worlds—their home ranges—got smaller too. Smaller animals have shorter GI tracts, so the average passage time—how long it takes from eating to pooping—was drastically reduced. Smaller animals have shorter lives. The average life span was cut by a third. In the Pleistocene, the average distance between eating and pooping was five and a half miles, or about a hundred and ten city blocks. After the die-off, it was down to a mile.

The large animals of South America were essential to the savannas and forests, spreading nutrients through their daily movements and seasonal migrations. They also spread seeds. You can thank the long-gone animals for your guacamole and dark chocolate; the plants likely coevolved with these mammals, relying on them to disperse their large seeds. The deep connection between plant producers and animal consumers was just as important on other continents, from Eurasia to Africa, North America, and Australia.

After the Pleistocene giants disappeared, phi declined by 98 percent, from a typical distance of about three miles down to just ninety feet per year. You can still see evidence of these pathways ten thousand years later in the Amazon, but they're starting to fade. Major human impacts on global biogeochemical cycles stretch back to well before the dawn of agriculture, the rise of the guano trade, fossil-fuel extraction, and nuclear bombs. Doughty and colleagues noted that some aspects of the Anthropocene, like blocked nutrient arteries and phosphorus limitation in the Amazon, might have started with these Pleistocene extinctions.

If the arrival of humans was like the onset of coronary disease to the animal circulatory system, the damage was debilitating but not fatal. You could easily mistake a rainforest for just a bunch of sticks, trunks, and leaves. But it would be unfair to dismiss the current suite of animals in the Amazon—monkeys, toucans, peccaries, tapirs, jaguars, leaf-cutter ants, and others—as irrelevant. (Areas without this suite of species have ammonia levels that are 90 percent lower than those with the animals.) Many species have gone extinct in South America. And there aren't all that many animals rustling about in the leaves. It might be the last place you'd want to look if you were setting out to prove that animals matter on an ecological scale. Until you look closely—and see the animal paths worn beneath the trees.

After the Pleistocene extinctions, there were global changes too. When the mammoths disappeared, almost thirteen thousand years ago, these huge, organic engines stopped grazing, putting an end to their methane-rich burps. The result: Fewer greenhouse gases, which led to temperatures plummeting ten degrees, which led to an ice age that lasted thirteen hundred years. We no longer worry about ice ages, of course, but the loss of animals puts us all at risk of losing critical ecosystem functions and services. To restore these systems, we have to help large animals, like bison, tapirs, and other native herbivores, return to the prairies, grasslands, and forests.

The trouble is, it's hard to repeat that model outside of national parks and game reserves. In Africa, many parks keep their wildlife fenced in. After massive poaching in the twentieth century, fewer than one out of ten savanna elephants survive. "I envy you your whales," Doughty told me. "There's no room left for big animals. Elephants still kill people."

MAKE A MUSCLE. It's likely that half of the nitrogen in your biceps was manufactured in an industrial plant. The nitrogen that forms your genetic code? Cooked up out of thin air. The RNA that translates the code and builds the proteins? Constructed from inorganic nitrogen, once inaccessible to animals unless it was captured by a plant, microbe, or lightning flash.

Peruvian and Chilean guano changed the world through economic and agricultural expansion. Some of the oceanic nutrients from the South Pacific might have ended up in clotted cream from Cornwall, merino wool from Australia, or a wheel of Parmigiano-Reggiano in Italy. But it wasn't enough. Nothing is ever enough.

The marine-derived nutrients fed people, but, in the form of nitrates, they also helped kill them. Potassium nitrate, or saltpeter, was an essential component of gunpowder, along with charcoal and sulfur. These last two ingredients were easy to find in the nineteenth and twentieth centuries, but potassium nitrate was relatively rare. Many gunpowder mills depended on guano for the oxidizing compound.

Guano merchants picked no sides; they sold to Germany, England, and the United States at the outset of World War I. But when a British naval blockade cut off supplies from Chile, Germany became desperate for new forms of nitrogen for explosives. In 1905, Fritz Haber, a physical chemist in Karlsruhe, had developed a new way of synthesizing ammonia, combining nitrogen from the air with hydrogen gas using high pressure and osmium, a rare catalyst. Osmium was too expensive for widespread use, and the application was limited until Carl Bosch, an industrial chemist, developed a way to mass-produce ammonia using iron, which was cheap and widely available, as a catalyst. Freshly armed with explosives, Germany and its allies kept fighting for another four years. As the war continued, Haber found ways to weaponize chlorine gas and other poisons and helped deploy them on the front in Belgium in April 1915. (His wife, also a chemist, committed suicide ten days afterward.) Three years later, Haber received the Nobel Prize in Chemistry.

The Haber-Bosch process revolutionized agriculture by turning nitrogen gas into ammonia for fertilizer.

$$N_2 + 3\,H_2 \rightarrow 2\,NH_3$$

British physicist Mark Sutton and colleagues call the Haber-Bosch process "the greatest single experiment in global geo-engineering that humans have ever made." The upside: Millions of acres of nutrient-depleted lands were kept fertile. Farmers typically apply about a hundred pounds of synthetic fertilizer per acre of active crops such as corn. (Organic farms don't use synthetic fertilizer, relying on crop rotation, raising crops and livestock together, or using organic manures and composts; production rates are often lower.)

We create more reactive forms of nitrogen like ammonia, about 165 million tons a year, than all natural processes combined. As a result, about half the world's human population is made possible by Haber-Bosch. It was a cornerstone of American agronomist Norman Borlaug's efforts to industrialize agriculture and reduce starvation. The head of the U.S. Agency for International Development called the new developments the "Green Revolution." Two years later, Borlaug won the Nobel Peace Prize.

But there were costs too. Millions of acres of marginally productive forests and prairies were lost beneath the plow, and billions of barrels of fossil fuels were burned to synthesize the ammonia. The

Green Revolution that sparked modern agriculture was driven, at least in part, by the shift from natural poop—manure, bones, and guano—to manufactured forms of nitrogen fertilizer. Sure, you can see the impacts of seabirds on Surtsey and bison in Yellowstone, but the effect of Haber-Bosch occurs on a much vaster scale. Atmospheric deposition of nitrogen has increased twentyfold in some areas, changing the nutrient balance in ecosystems accustomed to nitrogen limitation. This is perhaps most easily seen in coastal areas, where algal blooms have led to declines in water quality. Fertilizer runoff from midwestern farms and cities along the Mississippi River flows down to the Gulf of Mexico, causing blooms so large and intense they can cut off the supply of oxygen and form dead zones the size of Connecticut. Fish, marine mammals, and seabirds have moved on or died. Human fisheries dried up too.

Just as half of our nitrogen comes from labs, about 50 percent of the phosphorus we consume now comes from phosphate mines. Unlike nitrogen, phosphorus has no gaseous phase in the usual conditions found on Earth and it doesn't circulate in the atmosphere. It cannot be manufactured or destroyed. In coming decades—estimates range from thirty to three hundred years—phosphorus scarcity could threaten food production as ore deposits are depleted. Even if we never run out, phosphates might become so expensive that farmers can no longer afford them.

The Oulad Abdoun Basin in Morocco has an estimated twenty-six billion tons of phosphate, a marine sedimentary deposit that contains fish, shark, turtle, crocodile, and other vertebrate fossils collected over a period of about twenty-five million years. Bucket-wheel excavators, among the largest vehicles ever made, extract the rocks. Once the fossils are processed, the fertilizer is shipped and spread on fields and lawns, mostly in European countries. Phosphorus can be retained in the soil, enter crops and livestock, be recycled as manure, or—all too often—end up in streams, rivers, and other waterways as runoff, where it can cause harmful algal blooms. More than twenty-four million tons of phosphorus flows from fresh water into the oceans each year, and more than fifteen million are lost to erodible soils. A renewable resource has become nonrenewable, dependent on ancient sources rather than recycled ones.

Across the world, we've pushed the cycles of nitrogen and phosphorus over the edge of planetary boundaries, when irreversible, large-scale changes occur. We have become a geological force. Humans have transcended our role as agents of biological change—shredding food webs, plowing ecosystems, and annihilating biodiversity—to agents of geological change, directly altering the carbon, phosphorus, and nitrogen cycles. If you don't believe that animals can change the world, ecologist Joseph Bump argues, look no further than Haber-Bosch.

The animals are us.

6

Everybody Poops—and Dies

Aload of shit lumbered out in front of me. Like an armored dino-
saur, a soiled black tractor with flashing yellow lights pulled a
top-heavy manure spreader through the rain. Cow poop leaked over
the sides of the truck, covering the main street into our small town.

It was mid-May in rural Vermont. Mud season was over. The trees
were bright with light green leaves, the chlorophyll just ramping
up. The air was thick with the smell of cow poop: a little grass, a
little decay, a hint of sulfur. A swath of manure browned the edge
of a hayfield by the river. If you visit a dairy farm, you'll have no
doubt—given the smells, the movement of manure, the long-distance
shipping of milk and meat—that animals matter.

Our neighbors use tractors and fossil fuels to move the poop from
the dairy barn to the pastures and cornfields. On the edge of the
field, along the Winooski River and tucked behind some carefully
planted trees, the sewage-treatment plant works its magic. Life in
the Anthropocene is waste management—not just for cows, but for
us humans too.

Everybody poops. The simple truth is that defecation and urination, or pooping and peeing, are part of the daily rituals for almost all animals, the ellipses of ecology that flow through life. If animals are the circulatory system of the planet, then gastrointestinal tracts, from mouth to anus, are the pumps that keep the nutrients flowing. Much of the dirty work in the ecosystem inside our guts is run by microbes. Residing in the stomach, intestines, and colon, bacteria, fungi, and other microorganisms break down the complex carbohydrates, proteins, and fats that animals consume, making them available to the host. These microbes are us, forming a superorganism that is part animal, part traveling microbial circus. We can't survive without one another—though I'd place my bet on the microbes should the relationship sour.

"EAT, DEFECATE, REPEAT." With such a lede, how could "Hydrodynamics of Defecation"—a paper that appeared in the journal *Soft Matter* in 2017—not appeal? I rang up a couple of the authors.

"First off, what is soft matter?" I asked Patricia Yang. She had recently taken an academic position in her native Taiwan after completing a PhD at Georgia Tech and doing a postdoc at Stanford.

"A long time ago, there were only fluids and solids," she told me. Solid mechanics had its own techniques, fluid mechanics another. But later, practitioners in both fields came to the same conclusion: "Oh, there is something both of us cannot understand—something in between, like ketchup, toothpaste, and Play-Doh." At first, they called the field "polymer science," focusing on substances that acted like fluids—often big molecules with lots of small subunits, like DNA and proteins—but that was too limiting. The field itself was fluid—encompassing chemistry, biology, and engineering—so researchers came up with a new, looser term. "Okay, we give up," Yang said. "Those substances are soft stuff. Soft matter."

Yang and her academic adviser, mathematician David Hu, had a particular interest in the subject. They wanted to examine the fluid dynamics and biomechanics of feces as an extension of some earlier work they'd done on urination (we'll get to that in a minute). "For years," Hu and Yang noted, "animal trackers have been recording the shapes and sizes of the feces of a range of animals, yet there is no unified view of processes that generate feces." Bowel movements were judged qualitatively and subjectively by their frequency and appearance. There are several ways to classify those stools, including the Bristol Stool Chart for humans, from the hard lumps of constipation to the liquid consistency of severe diarrhea. The sausage is the gold standard, smooth or cracked.

Hu, Yang, and their colleagues wanted to quantify the forces, sizes, and amounts of feces produced by mammals. But where to get their samples? Yang went to Zoo Atlanta to observe and film animals defecating, and she watched a lot of videos of species that didn't live in the zoo. She collected measurements from elephants, pandas, lions, warthogs, and gorillas and from more housebroken species, like rabbits, cats, dogs, and humans. "I was especially taken with the hippo's rotator tail," she told me, much as I had been on my FaceTime safari in the Maasai Mara.

Yang noticed that many animals had cylindrical feces and that pooping took about the same amount of time, no matter the size of the animal. Elephants produce about fifteen pounds of poop a day—a hundred times more than a dog. How can it take the same amount of time to defecate? Elephants have rapid bowel movements—about

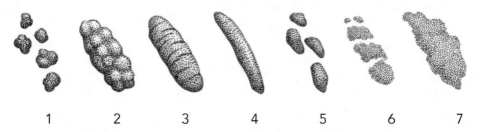

| 1 | 2 | 3 | 4 | 5 | 6 | 7 |

A typology of feces. The Bristol Stool Chart ranks feces according to viscosity, from hard constipation, 1, to liquid diarrhea, 7. Numbers 3 and 4 are considered normal, healthy shapes.

three inches per second, six times quicker than a dog's (humans average a little less than an inch per second). Yang started building a model.

"You have a process of transport of this feces column that's gotta be pushed out," Hu told me. He was at home, in a white T-shirt and black headphones. "The body already has some internal pressures and then you bear down, which increases the pressure..."

For a moment, he lost his way, gazing out the window from behind the computer screen. "Um, sorry, my wife is trying to pull our car out of the driveway. And she's going to yell at me because I parked it too close," he said with a catlike smile. "The driving forces are the pressure," Hu continued, "and the resistance forces are interactions with the walls around it."

The duration is the same for elephants, cats, and humans. "The magic number is twelve seconds," Hu said. At first, they were surprised, since the rectum of an elephant is ten feet long. They debated whether feces acted more like a fluid or a solid. There was no clear model to work from. "Materials have personalities," Hu said. "Molecules rearrange. Some of them have memory. If you take peanut butter out of the jar, it holds its shape. Ketchup flows a certain way. They have all these nuances. It's like a zoo of different kinds of fluids."

They used a rheometer, an instrument that gauges viscosity and elasticity. It's like an Oreo cookie, Hu said, with plates that measure torque, similar to the force needed to twist apart the two cookies from the filling. "Imagine doing that, but instead of filling, you have poop inside," he said. In the shared lab, there was a neighboring physicist's clean, non-poop-measuring rheometer that looked like a high-end espresso machine, and there was, well, the "Hu rheometer," stuck with the dirty work of measuring feces, mucus, saliva, and who knew what else. The soft matter.

While Yang was in the lab, she also recorded the smell and the buoyancy of the samples. Meat-eaters make sinkers and stinkers—their dense poop smells terrible. Think of your dog's poop bag, the stench of dog shit on your shoes, or the lingering smell of a litter box. Carnivores also poop less, since meat is easy to digest. Plant-eaters make floaters, and their fecal matter tends to be mild and rather earthy. Think of a well-ventilated horse barn or a cow pasture. The bison patties I picked up in Yellowstone had the scent of rich, earthy soil, which, essentially, they were.

After their debate about feces being fluid or solid, their measurements told them something new. "We realized we were going in the wrong direction," Yang said.

Poop moves like a solid plug, but more important was the thin layer that surrounded it: mucus, the lubricant that drives all the flow. The thickness varies with size. Larger animals have more feces and thicker mucus levels, allowing everyone, from cats to humans to elephants, to poop for about the same time. Without this layer, feces would be pushed out like toothpaste from a tube, and the forces needed would be tremendous. "You'd also get deformed feces," Hu noted, "which is not what your body wants." I thought of those golden stools at the center of the Bristol Chart. "Your body wants to cleanly get rid of it, so it has to provide some kind of lubrication on the outside."

Hu and Yang described the mechanics of defecation in their paper as something like a tugboat pushing a barge. It's Newton's first law of thermodynamics: objects at rest stay at rest, but under pressure from rectal muscles, fecal matter is pushed out of the body and onto the forest floor or into a litter box or toilet. Mucus helps the feces slide out. So defecation is a bit like a banana sliding out of its peel. Or perhaps a bowel movement is like a waterslide, giving new appreciation for potty talk: "I have to drop the kids off at the pool."

The hydrodynamics of defecation depends on the width and length of the feces, the diameter of the colon, and the amount of mucus sheared off the walls of the large intestine, which keeps everything moving along. Since mucus evaporates once it's released, it gave Yang a chance to measure it. She weighed the feces from animals as small as mice to as big as humans immediately after defecation and then followed the weight loss over time: after about thirty seconds, the mucus—about forty-five milligrams in a rabbit pellet—disappeared. There was a visual change too: the sheen came off the pellet. The bigger the animal, the larger the poop, the thicker the mucus layer. Yang and Hu later described their research as "a unified theory of pooping."

WHEN YANG FIRST arrived at Georgia Tech to work in Hu's lab, she didn't have a project in mind. "At the beginning," she told me, "the possible research topics were very diverse." As an applied mathematician, Hu, her major professor, had published on topics ranging from water striders to snake locomotion to the frequency that mammals use when they shake their fur to dry off. Where to start?

About the time Yang arrived, Hu had recently become a new father. He had been watching his infant son while changing his diaper. "I was really fascinated at how much pee little kids have," he said. Hu's son was about a tenth his size, but he peed for a long time. Hu wondered if animals might be the same. How do domestic cats compare to, say, elephants, which can outweigh them by a thousand times? It was time to do some fieldwork.

"One day David told me that there was a chance I could go to the zoo," Yang said, "but I'd have to watch how animals pee. 'Would you like to do it?'"

Yang told Hu, "If there's any chance I don't have to stay in this office, then I will go. And if there's a chance to be around animals, that's even better."

She scoured the literature for urethra measurements, flow rates, and bladder capacity, but her greatest resources weren't on Google Scholar. "Before we did the experiment at Zoo Atlanta," Yang said, "we looked at a lot of YouTube videos." Zoo visitors have a particular fondness for watching, recording, and posting animals peeing. "We got a lot of data from those videos," she said. They would later acknowledge these contributors. In between Zoo Atlanta and a National Science Foundation Young Career Award, they thanked demondragon115, ElMachoPrieto83, krazyboy35, MrTitanReign, relacsed, and fifteen other YouTube contributors.

Yang made the rounds of the elephant enclosures, but she really enjoyed the petting areas. "You have all these friendly, cute animals, like pigs and goats," she said. "You can just touch them and get closer to their butt when they pee."

Until they analyzed the data, Yang and Hu thought of the urethra as a pipe that connected an animal's bladder to the outside world. But they found that it actually accelerates the flow. "If you asked professors of engineering how long it would take for an elephant to empty its enormous bladder," she told *New Scientist*, "they would probably say half an hour."

Not quite. Elephants were great subjects. "They don't care if you are watching them," Yang told me. "And they have a regular schedule. Every morning at seven thirty, they pee at the same spot." No matter the time of day, all conversations stop when an elephant pees, because it's so loud. "A female elephant's urethra is a meter and a half long," Hu said. "And based on the width and the length, elephant pee comes out like five showerheads."

After hours of watching animals pee, Yang, Hu, and colleagues developed a model showing that the duration is the same for all animals. From six-pound cats to ten-thousand-pound elephants, the magic number is twenty-one seconds. Bigger animals have bigger bladders and longer urethras, which speeds up the whole process. Only small mammals, like bats and rats, break the rules, since they pee in droplets rather than jet streams. Urination for such small animals is a high-speed event, often lasting less than a second.

So there was a twelve-second rule for defecation and a twenty-one-second one for urination, but the variation was a lot higher for peeing. The twenty-one-second rule holds true because animals will urinate when the bladder is around two-thirds full. But behavior can influence the length of a pee. Dogs, for example, use urine as a scent marker, a way of communicating. They might not wait for their bladder to get full, taking shorter leaks.

Aging also plays a role, Hu said, since muscle tension in the bladder decreases in older animals. Some people can urinate for three to five minutes. As for those younger humans who dwell in the bathroom, the jig is up. That's more about the search for alone time or catching up on some reading than it is about the time it takes to defecate.

Once a week, the *New York Times Book Review* asks an author what books are on his or her nightstand. On my bathroom shelf: George Saunders's *A Swim in the Pond in the Rain*, Robert Sapolsky's *Primate's Memoir*, Peter Godfrey Smith's *Metazoa*, and Nicola Davies's *Poop: A Natural History of the Unmentionable*. Not that anyone asked.

In 2015, Yang and Hu were given an Ig Nobel Prize—a satirical award presented at Harvard that highlights real, if absurd, studies—after their work appeared in the prestigious *Proceedings of the National Academy of Sciences*. Hu wore a toilet seat around his neck during his acceptance speech. They would later receive a second Ig Nobel Prize for their work on the origin of the cubelike feces of wombats.

THERE ARE ABOUT seventy-nine million dogs in the United States and ninety-three million in Europe, all offering their human companions a direct view of the timing, shape, smell, and nutrient subsidies provided by animal pee and poop. They also shape the environment through nutrient dispersion, just like wild seabirds, whales, and salmon. Back before pooper-scooper laws were common, scientists found that dog defecations could elevate soil phosphorus levels in public parks for years after dogs had been banned from the area. In Belgium, scientists estimated that dogs contribute about a hundred and ten pounds of nitrogen per acre near walking paths of forests, grasslands, and other natural areas around cities, higher than what farmers typically apply to active croplands.

I've been able to watch these canine-derived nutrients delivered firsthand. Every morning at six thirty, I walk our dog, Zoey, down our quiet dead-end road. A wag of the tail is a great way to start the day—she's a restless ball of energy in the morning—but the first walk is also a mindful observation of defecation and urination. In summer, a leisurely stroll is fine. But in winter, it's all business. I don't want to be standing around in the Champlain Valley winter waiting for a poop or, worse, going out an hour later if we don't get a poopsicle on the first, cold try.

Following my chat with Hu and Yang, I had a new appreciation for Zoey's daily rituals. After a squat urination, she scratched the grass with her front and back legs, exposing the scent gland beneath her toes, like an alpha female wolf marking her territory in Yellowstone. Her pees were shorter than the twenty-one-second rule, about thirteen seconds each, I'd say, perhaps because we were walking the borders of her home range and she felt the need to distribute her urine a little farther, on the edges of lawns, margins of roads, in the town square, anywhere she might come across another dog.

When dogs are kept on leashes, concentrations of nitrogen in an area increase as the room to roam declines. Researchers in Finland found that total concentrations near trees, lampposts, and lawns along walking trails in Helsinki were much higher than they were on lawns with little dog traffic. And the chemical signature of said nitrogen revealed that dogs were its primary source. Walking trails are like Pacific salmon streams, well-traveled pathways for nitrogen delivery, in this case into urban green spaces.

OUR DOGS, FOR the most part, are trained to pee and poop on schedule. (Or perhaps they have trained us?) Jeremy Kiszka, a marine mammologist at Florida International University in Miami, wondered if we could use this attention to detail to answer some of the intractable questions about ocean defecation.

Let's face it, it's next to impossible to reliably collect whale poop in a typical field season. You've got the fog, the wind, the boat problems, the days without whales, and the days with plenty of whales but none of them pooping. We're lucky to get maybe twelve samples on the water over the course of a few weeks. And for the most part, we're restricted to fecal plumes—because we can see them. Whale pee is almost impossible to find in the field. It's like trying to tell when someone is peeing in a public pool, except in this case, it's a *really* big pool, often a little turbid, and maybe two thousand feet deep.

Kiszka thought he might have a workaround. He's no SeaWorld biologist; he's done fieldwork on sharks and whales off the Seychelles and Madagascar. He's also worked with whalers in Saint Vincent in the Caribbean to get biopsy samples for analysis of DNA and heavy metals. Wild whales, dead whales, and then he thought: *What if we could use captive animals to get at some of the intractable problems faced in the field?* Could captive bottlenose dolphins be trained to provide fecal and urine samples on demand?

Folks were wide-eyed at the prospect of potty-trained dolphins. "They shit like crazy," Kiszka told me as we drove down to Key Largo in his black Jeep one afternoon in late August. "It's awesome. We also trained them to urinate on command."

He had spent the past couple of months working with eight bottlenose dolphins and their caretakers at Island Dolphin Care, a small lagoon carved out of a canal in Key Largo.

As we walked barefoot out on the pontoon, I noticed a cute dolphin following me in the dappled water. My heart raced. With big dark eyes, bright pink smile, a few freckles, and smooth skin, Bella almost swept me off my feet. But I kept it professional and stayed on the deck. (The nonprofit facility has a swim-with-dolphins program.) Born in November 2000, Bella was one of Island Dolphin Care's stars. She still lived with her mother, Sarah, but she kept her eyes on me, almost flirting. I was smitten.

"Very seductive," Kiszka said, giving me the dirty eyeball. He had been boasting about their relationship in the car on our way to Key Largo.

Jealousy aside, we soon turned to the task at hand. The head trainer, Luke Bullen, called in the dolphins with hand gestures and a thin metal whistle. Sitting on the edge of the deck, he wore a navy-blue rash guard and sported a blond manbun. One at a time, the dolphins lined up, their backs resting on Bullen's curled-up feet, pink bellies to the sky. Liz Goetzl, the veterinary technician at Island Dolphin Care, carefully inserted a thin red catheter about six inches into Squirt's vent and pulled out a plug of feces that was a nice chocolate brown.

"Bounty time," Kiszka cheered from beneath his straw lifeguard hat, transferring the sample to a clear Eppendorf tube. Squirt got a couple of fish.

"Good girl," Bullen said with a flick of his wrist. "Go rinse."

Squirt swam off and circled back, belly to the sky.

"Let's see. Dry as a bone or Niagara?" Goetzl, wearing mirrored sunglasses and a light blue T-shirt that had all eight of the resident dolphins' names on it, wondered aloud. She put her fist gently on Squirt's bladder to ease the urination along, but there was little need. A yellow pool soon appeared around Squirt's genital slit. Goetzl drew the urine out with a small plastic syringe.

"Niagara! Good girl," she said with a big smile. We applauded, like you do for a kid being toilet-trained. Squirt got another fish as a reward, and Bullen sent her off with a hand gesture and the toss of a ring toy.

Goetzl handed the syringe to Kiszka, who transferred it to a brown plastic bottle to avoid sunlight. The process is efficient, friendly, and filled with rewards—vocal and culinary. The interns entertained the other dolphins, keeping them happy and distracted. If the dolphins minded this excretal roundup, they didn't show it.

Kiszka's graduate student whispered, "Dr. Dolittle science."

These samples would allow Kiszka to examine nutrient content—nitrogen, phosphorus, and iron—in the urine and the feces. He could also examine the gut-passage time—how many hours between a meal and a poop or pee. Until now, we have often depended on models from seals and sea lions to estimate the consumption, digestion, and metabolism of whales and dolphins. Kiszka was getting closer to understanding how cetacean nutrient ecology works. This matters because many dolphins, such as spinner dolphins off Hawaii, feed offshore and rest in shallow waters near the coast. The samples Kiszka and colleagues were collecting could provide a better estimate of the dolphin-derived nitrogen and phosphorus in coral reefs, seagrass meadows, plankton, and fish.

As we were on the pontoon processing the poop and pee, the dolphins were communicating through blows, clicks, and whistles in the lagoon. Individual dolphins have signature whistles, like human names, which they broadcast and respond to throughout their lives. Dolphins, like dogs, also use pee as a marker; they can detect urine samples from familiar individuals—friends or allies—ignoring or avoiding the unfamiliar or potential antagonists (which Bella and her lagoon mates wouldn't come across). Unlike dogs, dolphins can't smell; they don't have olfactory bulbs, so they likely use taste to detect friends. Echolocation allows them to explore the speed, size, and density of objects around them, and the taste of urine lets them know who's around, perhaps their reproductive status, and who knows what else.

Bella, Sarah, Squirt, and the others were doing something else when the trainers weren't looking. The trainers are very careful in monitoring the amount of food each dolphin gets. In addition to nutrient analysis, Kiszka did some DNA barcoding, using genetic markers to identify the prey species in the poop. They wanted to see if the DNA identified the squid, herring, sardines, and other fish that the dolphins were fed each day so they could use the same protocol for wild cetaceans. The expected animals were there—but so was a species not given to them in their feed. Had the samples been contaminated? It was a local snapper. Some dolphins had been grabbing a meal on the side, eating the fish that swam through the nets into the lagoon. (The mesh size keeps the dolphins in but lets snappers and other fish travel back and forth, at their own risk.)

The last sample collected, Kiszka thanked the caretakers and hurried off with six urine and five poop samples. It might have taken a few weeks to collect this many fecal samples in the wild, and we would never have gotten the urine without these captive dolphins and their trainers. Kiszka was in a rush. He had to get home to let his German shepherd, Wolfgang, out to pee.

"WE STARTED WITH the urination study just because I thought that the data was really interesting, and we could come in with some nice modeling," Hu said. "That study won us an Ig Nobel prize"—an accomplishment of which he was very proud—"but it introduced me to a lot of strange people." (When our conversation came around to my work with whale and dolphin defecation, I tried not to take the comment personally.)

One of Hu's new acquaintances, David Meyer, a Canadian civil engineer who studies piping systems, asked if he could help locate infectious-disease outbreaks in refugee camps and other high-risk areas by detecting new cases of diarrhea. They were interested in making an automated cholera detector for toilets, Hu told me. "If there's a high incidence of cholera," he said, "we wanted to send a signal so that the camp knows." The correct clinical treatment of cholera can be the difference between life and death; without proper care, cholera mortality in refugee camps can be as high as 60 percent. With appropriate clinical management, it can be well below 1 percent. Hu and his colleagues wanted to deploy a microphone to detect diarrhea assisted by machine learning. Changes in water, sanitation, and health can be hard to measure. But sounds can be anonymous, so it was easy to collect a lot of data. The noises could act as an early-warning system of an outbreak.

"What kind of noises?" I asked.

Urination is a jet, with the end of the urethra causing the urine to break into drops. "It has a tinkling sound," Hu said. Defecation is a series of discrete events associated with the speed and size of the object: plops. Flatulence has no fluid characteristics. There's a sphincter and there're flaps; the oscillation of the flaps generates the sound: "A fluttering flutter," as Hu described it.

They got the system to recognize the noises: Urination is a tinkle, defecation a plop, flatulence a flutter. (They used whoopee cushions to create the farts.)

Diarrhea is harder to describe; it's a combination of all three. The release of number 7 on the Bristol Stool Chart is close to instantaneous. Diarrhea doesn't follow the usual rules of fecal time, a wrinkle in the universal theory of pooping.

"We want the audio sensors to tell us which one it is," Hu said. So they got hundreds of minutes of experiments from YouTube. "There's a lot more diarrhea on the internet than regular poop and pee," Hu noted. "This DiarrheaGod on YouTube just drinks a bunch of colonoscopy prep and records himself," posting long videos of diarrhea noises. (The screen is black, but don't listen on a full stomach.)

Hu's colleague did a lot of his own self-experiments too, followed by apologies to his spouse and some cleaning supplies. Full disclosure: Some of my work on fasting whales occurred to me on the toilet, before a colonoscopy. The clear-liquid diet and laxative brought home that fasting whales probably have empty GI tracts for half the year.

Once they had the noises, they built a simulator for a toilet that could emulate the sounds. The next stage will be deploying the sensors in the field.

Hu's wide-ranging work—which has stretched from walking on water ("The hydrodynamics of water-strider locomotion") to sidewinding on sand to the physics of tossing fried rice (a combination of sliding along the wok and flinging into the air)—seemed to fly in the face of mathematician G. H. Hardy, who famously commented: "Is not the position of an ordinary applied mathematician in some ways a little pathetic? If he wants to be useful, he must work in a humdrum way, and he cannot give full play to his fancy even when he wishes to rise to the heights." Hardy should talk, since his Hardy-Weinberg equilibrium, a towering principle in population genetics, has been torturing biology undergrads for generations. (For his part, he thought the approach was very simple.)

Hu seemed to give full play to his fancy while also taking on practical questions such as how to prevent the spread of cholera and other infectious diseases.

WHETHER OR NOT Hu's diarrhea-detecting AI toilets materialize, the work he and Yang have done has been essential to understanding the mechanics of defecation. They, perhaps with another of the oddballs they met through the Ig Nobel, were the first to calculate the total amount of feces humans and our livestock—including both the animals we eat and those we use for transportation, like horses, donkeys, and mules—produce each year. It was cocktail-napkin science, as one of Hu's colleagues described it, launched over a beer, with absolutely no funding behind it. (Not an unusual situation for sometime scatologists.)

We mammals defecate about 1 percent of our body weight every day. Every three or four months, an elephant poops an elephantine amount of poop. The same holds true for dogs, hippos, bears, and, presumably, whales. By my humble calculations, that means we humans poop above our body weight, flushing more than three of ourselves down the toilet (or into the outhouse or wherever) each year. When you add it all up, domestic animals—primarily cattle, chickens, and sheep—produce about eight trillion pounds of poop per year. Depending on how you stack it, that's the weight of ten thousand Empire State Buildings or seven hundred Great Pyramids of Giza. (Pyramids fit the silhouette of a poop emoji much better.) We humans make about two trillion pounds each year.

We are animals too, of course, and there's no reason we couldn't play a more integral role in the nitrogen and phosphorus cycles instead of manufacturing and mining to replace the nutrients that we let leak away. Humans did this for millennia before we broke the loop; consider the simple outhouse, the arborloo (a movable pit that will later fertilize a tree), or composting toilets that use sawdust and air to enhance decomposition.

With the loss of megafauna and the near-death of phi—Doughty's dispersion measure for dung—humans created a paradox. Farms and lawns, subject to harvest and runoff, tend to lose nutrients. Rivers, lakes, and coastlines downstream of agricultural land and urban areas have too much nitrogen and phosphorus. The water bodies are subject to algal blooms, which are low in oxygen and can cause widespread die-offs of fish and invertebrates.

One August afternoon when cyanobacterial blooms in Lake Champlain were in the news, I stopped by the Rich Earth Institute (REI) in Brattleboro, Vermont. On the way in, I passed a small exhibit of zero-flush, eco-flush, and composting toilets in gleaming white. Kim Nace, Rich Earth's cofounder, and Julia Cavicchi, the education director, gave me a tour of the research center. Their mission: to advance the use of human waste as a resource—peecycling. Nace, who started Rich Earth in her basement in 2012, had a warm smile, rose-rimmed glasses, gray hair, and a nose stud. REI has since grown into a large lab in a low-slung building on the industrial outskirts of Brattleboro, not far from the solid-waste management district. The Rich Earth Institute has drawn attention from media outlets, including CBS, PBS, *The New Yorker*, and the *New York Times*.

"Once it was in the *New York Times*," Nace said, "there were hundreds of comments. And many of them were like 'I've been doing this for years.'" Suddenly, people felt free to admit that they had been doing their own form of peecycling for decades. "The *New York Times* somehow made this legitimate to discuss in public," Nace said. The institute has about two hundred urine donors who deliver their "bodily nutrients" to REI. Others use their urine to fertilize their own gardens.

Think before you flush: Many of the nutrients released into sewers go straight through wastewater-treatment systems out into bays, lakes, and drinking-water supplies. Harmful algal blooms, driven by too much phosphorus, can cause serious illnesses, such as paralytic shellfish poisoning, and kill fish, wildlife, and pets. When people started shifting night soil (poop and pee removed under cover of dark) away from farms to rudimentary treatment centers, they broke the human phosphorus cycle, "reshaping its loop into a one-way pipe," as Julia Rosen wrote in the *Atlantic*.

The Rich Earth Institute was determined to restore the loop: Eat. Urinate. Sanitize. Fertilize. Grow. Recycling pee—"liquid gold," as REI describes it—could cut down on mining and the energy-intensive Haber-Bosch process and recover more than thirteen billion dollars' worth of nutrients per year. There's enough nitrogen, phosphorus, and potassium in urban wastewater to offset more than 13 percent of the demand for agricultural fertilizer and enough energy, in the form of ammonia and other molecules, embedded in wastewater to provide electricity to 158 million households. We could even recycle pee at a wastewater-treatment plant and distribute nutrients to local farms, a new twist on the locavore movement.

Occasionally, as the three of us walked around Rich Earth, we could smell urine, which was being processed for release.

"Do you want a tomato from our urine-fertilized garden?" Tatiana Schreiber, the social research director, asked, handing me a Sun Gold tomato.

Did I hesitate for a moment? "Sure."

It tasted as bright as it sounds.

Their garden, on the industrial edge of Brattleboro, looked far better than ours at home—the urine-fertilized tomatoes more numerous, the eggplants bigger, the corn taller.

The Rich Earth Institute works locally and more broadly to bring urine recycling to other towns and cities. "I'm a big fan of container-based sanitation," Nace noted. "One of the problems with centralized treatment plants is that when you have a flood or combined-sewer overflow, you're just dumping tremendous amounts of waste into the water." Shores, lakes, and rivers receive the sewage in a big pulse. Container-based sanitation, in contrast, is modular and easily manipulated. Or so Nace claimed.

On my way out, I eyed some translucent containers. With a bright red funnel and a two-and-a-half-gallon collection jug, each of these peecyclers resembled a Duchamp Readymade. Cavicchi told me to pick my favorite from a box of plastic balls, commonly used to fill ball pits in indoor playgrounds.

I took one of the portable urinals home, and for a while, it decorated the stairs to my basement office. I thought it would add a sustainable Dada touch to the bathroom. But my family was firmly opposed. They weren't interested in the peecycler, modern art or not, and forbade me to use it in the full bathroom, the half bath, or my office.

I protested, echoing a talk I had heard by Chelsea Wald, author of *Pipe Dreams: The Urgent Global Quest to Transform the Toilet.* Urine diversion has a long history, I told them, stretching back thousands of years to when ancient people used urine as fertilizer and medicine. The invention of the modern toilet, that sanitation revolution, put a stop to the practice, shunting most pee-borne nutrients underground or out to sea. If people used urine as fertilizer, I said, we could cut global nitrogen production by a quarter. We could each save about four thousand gallons of water per year (from avoided flushing), reduce nutrient pollution by about 50 percent, and lower greenhouse-gas emissions by reducing the energy needed for conventional wastewater treatment.

They were fine with the sanitation revolution, thank you very much, my family informed me. They wanted no part of the peecycler in our home. Outside, maybe, but that seemed a little risky and risqué with the neighbors. So I got on the bus with my Readymade (empty—come on) and took it to my rental in Cambridge, where I was doing a fellowship. Now I had my own personal peecycler. In honor of Hennig Brandt, who supposedly recruited beer drinkers to supply the ingredients for his seventeenth-century experiments, I drank a couple of craft brews as a primer. It took about two weeks to fill the jug.

There are other benefits to recycling your pee. There's the warm glow—an emotional one; please—the feel-good effect of green behavior that's been recorded in the sustainability literature. These intrinsic rewards come with efforts to heal rather than harm the Earth. At REI, the goal is to move us in this direction.

After I filled up my container, it started to bronze with age. Over the winter, it fermented in the basement (considered a good thing at REI, since that can help sanitize it). What to do with this valuable fertilizer? My family stuck to its hard no when I mentioned applying it to the vegetable garden. My friends said they weren't growing vegetables that year. And the Rich Earth Institute has strict policies against spreading urine without permission.

So one spring afternoon, I dropped my urine off at REI's pee-cycling center in downtown Brattleboro. It would go into the cycle: Eat. Urinate. Sanitize. Fertilize. Grow. Nace and her colleagues would pasteurize the urine following World Health Organization guidelines and then use it as a supplement for nearby farms. Much of the urine—about a thousand gallons per acre, the annual production of eight people—goes to hayfields, replacing synthetic fertilizer. (Take that, Haber and Bosch.)

The resistance to pee fertilizer, I realized as I drove back from the Brattleboro drop-off, was in line with how we humans hide the rest of our sewage—and dead bodies.

POOP AND PEE are often flushed down the toilet and treated in septic systems or piped into the ocean. That's the daily flow, but there is also a final pulse: our dead bodies are boxed up, burned up, buried, or walled in, often at great environmental costs involving high carbon emissions and other pollutants.

Can we join in on the decomposing and return our nutrients to nature? After years of studies at the Forensic Osteology Research Station of Western Carolina University—also known as "the body farm"—researchers have optimized human composting, from death to soil. Recompose, a green funeral service in Seattle, places bodies in a capsule filled with wood chips, alfalfa, and straw. Gases such as cadaverine and putrescine, common in putrefaction, the fifth stage of death, are treated with a biofilter before they are released. The bones are ground into powder. It takes about four to six weeks for a body to decompose fully.

We have the technology, but "with death, change comes slowly," Caitlin Doughty, a mortician and advocate for funeral reform, warns. Only a few states, among them California, Washington, and Vermont, have passed legislation that legalizes human composting.

What to do with this human-compost soil? It can be scattered in a cemetery, placed in a grave, or given to the family to use in a garden. But why not think outside the box? At Recompose, you can donate your remains, a cubic yard of soil, to Bells Mountain Conservation Forest in Washington State. Saplings are planted with human compost to help shade streams, which could restore habitats for spawning salmon and steelhead. Human remains could replace the ghosts of salmon carcasses.

7

Beach Read

If you're reading this book on a tropical beach, consider yourself lucky—especially if a magnificent frigate bird shits into your open book, as one did on my copy of Peter Matthiessen's *Birds of Heaven* a few years ago on the Dry Tortugas off southern Florida.

Plunge your hand into the rough sand, the colors of oatmeal, cream, sandstone, slate, that brought you there. You really can see the biosphere in a grain of sand—from the flying fish to the frigate bird, the coconut palm to the elkhorn coral and the parrotfish—on this island beach. Most people don't realize that when they stretch out on a beach in Hawaii, they're lying on a bed of animal waste. Biogenic sand, as it's known, can pass through the gut of a parrot-fish or come from ground-up animals—sponge spicules and barnacle fragments—and coralline algae.

I slipped into the waters off Kona on the Big Island of Hawaii to find the sand's source. The reef showed signs of wear and tear, with lots of dings in the corals from careless snorkelers, but there were large fish, as bright as Broadway, all around. The uhu, as the

Hawaiians call parrotfish, were hard at work. As I snorkeled along, I heard a hard crunch. Parrotfish can bite through concrete, and as I followed them, it looked like they were kissing the reef, vampires sinking their fangs into stone.

The first uhu I saw was a looker, a spectacled parrotfish in Chihuly blues and greens that seemed to be lit from within. With an overbite of a thousand teeth cemented together in a beak formation, it bit through the coral skeletons like Tootsie Pops. Parrotfish can crush calcium carbonate—also known as limestone—and release it through their poop and gill slits as sand and silt. They do more than just move nutrients; they are physical engineers, biting and scraping dead corals and grazing on rock surfaces to get at large seaweeds, small algae, and even microbes.

The parrotfish I was following released a trail of white sand behind its bright orange fins. The poop held together for a moment, almost fabric-like, before settling to the seafloor. Meanwhile, a nearby parrotfish expelled a loose puff of silt from its gill slits, giving the impression of a strange locomotive. I passed my hand through another parrotfish plume; it dissolved into a thousand grains.

If you follow an animal long enough, it will poop. So I followed another favorite—they're all my favorites—a star-eyed parrotfish, or pōnuhunuhu in Hawaiian, along the edge of the reef. It was a short wait. This particular male, with pink stars around its eyes and a neon blue body, was eating and pooping almost simultaneously. It's no surprise. Elsewhere in the Pacific, humphead parrotfish take about three bites of dead corals per minute. They poop about twenty-two times per hour!

Parrotfish play a crucial role in coastal ecosystems. Beach-builders, they use their formidable dentition, dental plates derived from fused teeth, to feed on corals, seaweed, rocks, and sand. The algae, along with the bits of rock, coral, and sand they consume, is ground up by small teeth in the throat, known as the pharyngeal mill, making it more digestible. In the process, limestone fragments are turned into sand, and sand into finer sand. Four out of five grains of sand on many tropical beaches are poop from parrotfish meals. By their constant grazing, parrotfish protect the reef, helping to keep corals clear of overgrowth by seaweeds and reducing invasive species. As they scrape, they open new areas for young corals and other bottom-dwellers to settle and grow.

A single green humphead parrotfish, native to the Indian and Pacific Oceans and measuring more than four feet long, can poop almost ten thousand pounds of sand annually. Hire that fish! Three of them can deliver a concrete mixer's worth of sand to the beach each year. The bigger the chomp, the more living and dead corals they consume, and the more beach sand they excrete. I looked at the creamy rim of beach along the island's edge. The parrotfish I was following were smaller than humpheads, but they could still generate up to a thousand pounds of sand—twenty-five sandbags' worth—per year. I was watching them slowly building the beach of the Big Island, one poop at a time.

Nick Graham, the British ecologist who studied the widespread impacts of rats and seabirds in the Chagos Archipelago in the Indian Ocean, has recently turned his attention to parrotfish. Graham has shown that these herbivores grow larger in the waters around seabird islands. Why? The answer will be familiar by now. When terns, boobies, noddies, and frigate birds poop, they release nitrogen and phosphorus that is picked up by seaweeds, plankton, corals—and, eventually, fish. Parrotfish grow larger with higher-quality food. Larger fish have bigger bites, and grazing rates are three or four times higher on seabird islands. The bigger these beach-builders get, the better they are at reducing seaweed, opening areas for young corals, and pooping vast quantities of beautiful tropical sand.

"These seabirds," Graham said, "might be able to maintain island growth in the face of sea-level rise." Maintaining ecological pathways from ridge to reef could be key to saving low-lying atolls in the coming decades. With the help of the birds, parrotfish form a neon firewall against the reef's demise.

***

THE NEXT DAY, I drove down to Kealakekua Bay. The waters were clearer, the reef far more colorful. At times, I felt like I was snorkeling through an aquarium. I saw three of the island's native uhu species: the star-eyed, spectacled, and bullethead parrotfish. I could hear their biting sounds, like hammers crushing rock. Noisy reefs are healthy reefs, ringing with the continuous popping of snapping shrimp, the grunts and howls of fish, and, when you lift your head above the water, the cries of seabirds. The racket indicates a vibrant marine community, helping fish and invertebrates in the open ocean locate living reefs.

Even in this seemingly well-protected area, I noticed that the parrotfish were skittish. I later learned that local hunters speared the fish while they were sleeping on the reefs. No wonder the survivors swam off when I approached. One biologist told me he was shocked by the scarcity of parrotfish when he moved back to Hawaii in 2013. Where they had been abundant in the 1970s, they were now surprisingly rare. Their behavior had changed too—rather than traveling in large schools, parrotfish now tended to be solitary. On Oahu, only about one in twenty are left. This devastating 95 percent decline is a result of commercial and recreational fishing, especially nighttime spearfishing, and pollution from sewage. If parrotfish disappear, they won't be able to perform their ecological role—building beaches and protecting corals.

There were plenty of other lookers, including the state fish: the humuhumunukunukuapua'a, or reef triggerfish. The drop-dead-gorgeous orange-spine unicorn fish was also present, as were the reticulated butterfly fish, Moorish idol, and gold-rim surgeonfish. It turns out that many of these species play important roles on the reefs. There is evidence that parrotfish, butterfly fish, and other coral-eaters help spread the photosynthetic microbes essential for coral survival. They ingest zooxanthellae, the algal symbionts that grow in the tissues of corals, when they feed. Their fecal matter is loaded with these microscopic algae in densities that are several orders of magnitude higher than in the surrounding seawater. So as the fish move across the reef, they spread the essential symbionts to other corals, a potential benefit to young settling polyps—as individual corals are known—that haven't received these microbes yet. The zooxanthellae can benefit, too, by being dispersed to new reefs.

There is more to the community than just these showy species. The modest sponge—a sedentary, ancient, and rather simple creature—plays an outsize role in reef productivity. Rob Toonen at the University of Hawaii later explained it to me: "Why are coral reefs—these incredibly biodiverse and productive systems—found in these incredibly low-nutrient waters? For a long time, people have thought about the corals as the photosynthetic engine of the reefs. But not very much eats coral, and corals don't produce an awful lot of food." The missing link was the sponge, often clustered in large gardens. These filter feeders, many with glass skeletons, take organic matter dissolved in the water column and turn it into fecal particles that feed other marine invertebrates, such as amphipods, copepods, and polychaetes, animals that are at the foundation of coral-reef diversity. "We're starting to realize that the nutrients on the reefs have very little to do with the corals," Toonen said, "and an awful lot to do with sponge poop." Often overlooked, even this simplest, most stationary of organisms can transform the reef. It's an ancient, silent pump.

ONE AFTERNOON, I swam out from one of Oahu's most beautiful beaches. Green volcanic peaks curtained the horizon; a near full moon rose to the east. The soft warm sand, a mix of biogenic and basalt grains, had been groomed smooth. There was even some microplastic mixed in. (Just about every beach has a little plastic these days.)

With me were two of the world's experts on reef fish: Mark Hixon, a reef ecologist at the University of Hawaii, and Brian Bowen, a biogeographer with a specialty in coral-reef fish and molecular genetics. (Before going to UH, Bowen was my mentor at the University of Florida.) We snorkeled through Hanauma Bay, an old volcanic crater, the remains of an eruption that occurred about thirty thousand years ago, well before humans arrived. It's a popular swimming and snorkeling spot, not far from Honolulu. On Tuesdays, the place is closed, and the fish get a break. Hixon, who runs a research project there, invited us out to join him. He cautioned us to walk with care down to the water so as not to disturb the carefully groomed sands. Even a beach needs a day of rest.

When he was younger, Bowen was a killer of parrotfish, hunting them for their DNA. But eventually, he lost his taste for the work and stopped doing what scientists euphemistically call *invasive sampling*. I had also been complicit in the death of a parrotfish. A few years ago, my colleagues and I ordered a meal on a beach in Bahia, Brazil, that included stewed fish. It was a lovely white meat in a rich tomato sauce. At the end of the meal, a colleague asked our server what kind of fish was in the stew, and we learned, much to our horror, that we had eaten parrotfish. (Many of the carnivorous reef fish along the coast had already been hunted out.) The restaurant was perched on a sandy peninsula that was probably built and maintained by the parrotfish. We had consumed a local ecosystem engineer, and quite possibly an endangered one: the greenback parrotfish.

Fishing is forbidden in Hanauma Bay, which is designated a state underwater park, and the parrotfish there seemed less skittish. I snorkeled over to an enormous spectacled parrotfish, the largest uhu I had ever seen. It cracked into the corals and rocks below as if taking a bite out of concrete.

Having worked in the Caribbean, where overfishing has decimated fish populations, I was astonished by the color and abundance of fish on Hawaii's reefs. Coral reefs are the premier hot spots in the oceans; about a million species—a quarter of all marine animals and plants—spend at least part of their lives on these complex habitats. We saw butterfly fish, elegant coris wrasses, boxfish, and puffer fish in a forty-minute snorkel. The reefs also protect coasts from storms and in many areas serve as the intersection between the deep sea and shallow ecosystems like seagrass meadows and mangrove forests.

The weather started getting rough, so we snorkeled back toward the beach. There was another huge parrotfish swimming over the sand. What was he doing in that shallow water? Making a deposit, perhaps, or, more likely, checking on his harem. He chased a younger female back into the deep.

All the females were younger. Parrotfish are hermaphrodites, undergoing a sex change from female to male after a few years, depending on population density and growth rates. During this transition, the local uhu shift from a rather muted gray and red to eye-candy greens and blues with bright pink, yellow, and orange patterns from beak to tail. A successful male might have a harem of two to five females that he presides over, protecting a territory of maybe three thousand square feet—bigger than a typical American home. When the male dies, the largest female from the harem typically moves on to the terminal male phase and takes over. Occasionally, a fish will change sex without the showy colors. These sneaker males, as they're known, can travel with the females and mate without drawing attention from the bigger, showier males.

The visibility was getting worse, and Hixon apologized for the murkiness. Bowen responded: "Any snorkel, like any dive, is a good one if you come back alive." We retreated to the Kona Brewing Company to talk parrotfish. A long-necked bird came over to our table, looking for scraps. An egret walks into a bar...

The *Kumulipo*, the Native Hawaiian creation chant, says the world began with a single coral polyp, Hixon told us. He wore a navy-blue baseball cap and a University of Hawaii T-shirt; his woolly eyebrows almost perched like caterpillars over his Maui Jim sunglasses. "And I think that's really cool, because it's exactly how it begins," Hixon said. "A volcano hits the surface, coral larvae settle, and a reef starts. I love that."

I did too, since the story sounded familiar. Darwin wasn't too far behind the Hawaiians in understanding the development of reefs. He intuited that coral atolls, ring-shaped reefs, were the remnants of volcanoes. Corals settle and grow along the shallow basaltic rock of a new volcanic island or seamount. As the island ages, the volcanic rock erodes, and the seamount disappears. Life takes over—a ring of reefs and sand built by corals, sponges, parrotfish, and seabirds. In many places, these atolls are all that remain of long-forgotten volcanic islands.

Coral reefs, as Alfred Russel Wallace knew back in the 1800s, are marine-animal forests. Each polyp is like an upside-down jellyfish cupped in a limestone skeleton of its own creation. Much of this book focuses on how animals shape ecosystems that are dominated by plants—trees, grasslands, kelp forests—but in this case, the animals provide the seascape themselves. Sponges, sea pens, and tube worms can also form marine forests.

Corals are like ocean farmers. Related to jellyfish, hydras, and anemones, they provide shelter and deliver carbon dioxide to their symbiotic algae, and in return, the zooxanthellae synthesize energy-rich carbohydrates for their hosts. Corals aren't the only invertebrate farmers; it's a relationship that has arisen in the sea several times—many species of clams, anemones, and sponges have algal symbionts living in their cells.

"Herbivores—the fishes that eat seaweeds—are extremely important on coral reefs," Hixon told me, "because they keep the reef surfaces clean of seaweeds so that corals can grow." Seaweeds grow faster than corals, competing for light, space, and nutrients. In some cases, they will smother corals to death, especially when runoff from land provides them with too many nutrients, like human feces and agricultural fertilizers. Herbivorous fish are like gardeners, weeding the reef, each using a different technique or preferring a particular type of algae. "You need different garden tools to cut the grass and trim the hedges," Hixon said. Surgeonfish nibble seaweed off the surface of rocks and corals. "Parrotfish are the heavy lifters of the herbivores. I would call them the lawn mowers of the reef."

With their fused beak-like teeth, uhu scrape and excavate the reef, releasing nutrients from dead corals and algae into the water. "The scars they leave behind are where the baby corals grow," Hixon said.

The parrotfish open the space, and other fish eat the seaweed. In a positive feedback loop, the coral reefs create shelter for the herbivores. In addition to consuming seaweeds and opening up areas for new corals, parrotfish provide nutrients through their feces, turning a coral competitor into a coral fertilizer. "Corals thrive where there are lots of fish," Hixon said. "The fish feces and urine fertilize the corals, which have single-celled plants living in them. A healthy reef has all these interconnections," Hixon added. "Thank goodness for uhu."

MY ROOM AT the Lanai Suites on Coconut Island, the home of the Hawai'i Institute of Marine Biology, had a stunning view of He'eia, a steep green mountain looming over Kaneohe Bay. I could follow the steep green furrows of the mountain down to the blue waters, dark coral reefs, and white sand just outside my window.

Those clear waters were too tempting. I walked down to the sand at the end of a shortened workday and put on my mask and fins. (My host, Brian Bowen, told me it was fine if I swam alone: "Just don't die.") I snorkeled along the shallow reef edge, admiring the abundant small fishes—wrasses and butterfly fish, to my untrained eye. The corals looked healthy. They were dominated by two modest but resilient species: rice corals, with small beige polyps that resembled an old shag carpet at a certain angle and in a certain light, and finger corals, stout digits clustered in yellowish mounds. The bay was far from pristine—cargo and fighter planes buzzed overhead and did touch-and-goes on Marine Corps Base Hawaii, and much of Coconut Island is lined with concrete and nonnative mangroves—but the fish and corals seemed to be thriving.

When I told Bowen about my swim, he said I would have had little incentive to get in the water a few decades earlier. Between the 1940s and 1970s, the reefs had mostly disappeared. "The sewage used to go right into the southern end of the bay," Bowen told me, "and it could be weeks before it flushed out to the sea." Light penetration declined, nutrient levels climbed, and there were new dredging activities in the area. The bay, once described as a coral garden, was showing signs of stress.

I walked up the hill to chat with Bowen's coconspirator at the ToBo lab, Rob Toonen. Their lab had published dozens of papers on the genetics of the reef fish of Hawaii and marine organisms around the world, but I wanted to talk to Toonen about his work on the unnatural history of Kaneohe Bay. When Toonen arrived at Coconut Island in 2003, the reefs looked pretty good. "I would have never known that there wasn't a coral reef here if I had not interacted with our community," he said, "and found out the oral history from our *kūpuna* in the area." In Hawaiian, *kūpuna* means "ancestor or honored elder," the keepers of traditional knowledge and local history. For centuries, Hawaii was managed with an *ahupuaʻa* approach, which acknowledged the connections between the mountaintops and the sea and integrated terrestrial, freshwater, and marine ecosystems.

In the twentieth century, the traditional practices of the land—such as fishponds and terraces to collect stream runoff for "canoe plants" brought over by the Polynesians, like coconuts, yams, breadfruit, and taro—started to decline. Kaneohe Bay suffered decades of insults: a military base with massive dredging; livestock grazing, erosion, and sediment loading; and overfishing. Sewage from cesspools, septic tanks, and Kaneohe Marine Corps Air Station were discharged into the bay. Nitrogen and other nutrient levels went up, visibility levels went down, and algae thrived. Before the Western Era, corals covered 80 percent of the bay's floor. The number declined to about 60 percent as agricultural sediments flowed into the bay. Corals almost disappeared in the 1970s after land was developed and sewage flowed in.

According to Toonen, the bay was like pea soup in the seventies. There were major sewage spills. The reefs weren't functional; there were too many nutrients—causing algae to overtake the ecosystem—and not enough light penetration for corals to thrive.

Without corals, the reef fish—parrotfish, surgeonfish, and butterfly fish—also declined. Livestock grazing caused deforestation, erosion, and sediment loads in nearby streams and, eventually, the bay. The classic profile of Coconut Island is known to people of a certain age from the opening credits of *Gilligan's Island*, and it occurred to me that at the time it was filmed, the island was surrounded by water filled with sewage.

The reefs got hammered in those years, and so did Hawaiian culture. Toonen suggested that I chat with Kawika Winter, a professor of biocultural ecology at the University of Hawaii. Coconut Island is small, so I walked under the palms to Winter's office.

"In the 1950s and '60s, to be Hawaiian meant that you were ignorant," Winter told me. We were sitting in a common room outside his office; he wore an aloha shirt, shorts, and flip-flops. "You know all the stereotypes: lazy, low IQ, and prone to crime. There was no pride in being Hawaiian." In the seventies, that started to change. "The Hawaiian renaissance was inspired by the American Indian movement and the occupation of Alcatraz [Island]," Winter said. "A bunch of people who would've been my parents' age—I was born in '76 and that was right when all this was happening—they were starting to occupy islands."

There was a growing sense of Indigenous agency, Winter has written, founded on the concept of aloha *'āina*, or love of the land. During the 1970s, after decades of "pimping out Hawaiian culture to bring in tourists," Winter said, there was a revival of Hawaiian language, arts, culture, philosophy, and spirituality. "There was this great resurgence of pride in being Hawaiian, including traditional and cultural practices of stewarding the land." Among these practices was the concept of *kapu*, which restricted fishing at certain times of the year (violators faced extremely harsh penalties, especially around spawning season) and encouraged sharing among communities.

By 1974, there was increased pressure to reduce the human waste in the bay. Along with the promotion of traditional Hawaiian practices of managing ecological systems from ridges to reefs, there were federal changes in how water pollution was handled. In 1979, sewage was diverted to deeper waters offshore. "You see the bay start to clear up," Toonen said, "and you see a change in the community structure." Nutrient levels, turbidity (murkiness), and the numbers of phytoplankton and other larger algae all went down, creating much better circumstances for the corals that remained. Coral cover increased from a low of 10 percent of the bay to 60 percent. Limiting dredging and protecting the local fishes also improved the bay's health, so much so that young scientists today have no idea how bad it was a mere fifty years earlier.

The recovery of the Hawaiian reefs is one thing; restoring native forests is quite another. In the water, I was struck by the abundance of colorful reef fish and blue, pink, and purple corals. But when I hiked in the Hawaiian forests, there were few flashes of color, and it was quiet except for the calls of a few nonnative birds, such as the white-rumped shama, a melodious echo in an empty forest. Pigs might have arrived with the Polynesians. Rats showed up in waves from other islands in the Pacific and with Europeans too. No matter their pedigree, they destroyed habitat, preyed on birds, and enhanced the spread of mosquitoes through the islands. Hawaii has lost two-thirds of its native birds since humans arrived, about sixteen hundred years ago, many of them disappearing after the introduction of avian malaria, likely from larval mosquitoes in the drinking water of whaling ships. "Ancient Hawaii is a ghost that haunts the hills," E. O. Wilson wrote in *The Future of Life*, "and our planet is poorer for its sad retreat."

"Native Hawaiian wisdom teaches us that changes in the land have impacts on the ocean," Kawika Winter noted, "but how does the health of the ocean affect the health of the land? That connection for me is seabirds. Birds once blackened the skies of these islands, and when they did, they deposited lots of nutrients up in the forest." The forests coevolved with this marine-based fertilizer, which, according to Winter, was "like straight shots into the root system." These seabirds—like the ʻuaʻu (the Hawaiian petrel) and the

ʻaʻo (Newell's shearwater)—often burrow at the base of trees, but when the birds disappeared, forests were overrun by invasive species. "Our only native forest left is at the top of the mountains," Winter said, but the invasives keep spreading uphill as the climate warms. The native trees aren't as resilient and strong as they should be "because they're not getting the doses of guano fertilizer that they co-evolved with," he noted. "It's no surprise that our forest is in trouble."

People think you can cut down trees and replant native species, but it doesn't always work that way, Winter pointed out. There is the issue of lost nutrient regimes, and the mycorrhizae, the fungal symbionts around the roots of trees, might be damaged or destroyed. "What are the hidden elements of the system that you need to restore so that the visual parts of the system can survive?" Winter asked. "I see birds as a big part of that story."

It is not just the nutrient transfer that is lost but the cultural connections too. "Hawaiian creation chants," Winter noted, "account for the birthing of the islands, the biodiversity of its lands and seas, and the Hawaiian people." They are all *kūpuna*, elders and ancestors, with kinship extending beyond humans to the flora and fauna. Hawaiians still sing about birds that are now extinct, because they are retained in the old chants and hulas. "We sing these songs," Winter said, "and nobody knows what these birds even looked like.... Even our elders don't know what these birds sound like, they've been gone so long."

Hawaiian culture is dynamic and evolving, like all cultures on the planet. "Our language is alive, and people are writing new songs in Hawaiian, and they don't talk about the birds we've lost," Kawika Winter said. He calls these *coextinctions*, and rightly so. Just as the loss of an animal can mean the loss of its parasites, predators, and ecosystem function, the loss of one of Hawaii's birds is the loss of culture: stories, songs, traditions.

"There's this one hula that somebody composed, a traditional dance, that's about the coqui frog," a species accidentally introduced to Hawaii from Puerto Rico in the 1980s. "Of course, it makes sense," Winter said. "We create art about the systems around us. We are now making dances and songs about coqui frogs because the biodiversity and forest where our ancestors made songs isn't there anymore. We've had extinctions of those songs and ideas. All species have lessons to teach, and they're encoded in our language and the songs. But those lessons are disappearing."

Federal guidelines and local aloha *'aina* helped restore Kaneohe Bay. A focus on restoring wildlife and reducing the impact of invasive species, like feral pigs, could do the same for the lands across Hawaii. Winter has been working with local managers and hunters to create areas that are pig-free, which benefits the native animals and plants, and in areas where hunting is encouraged, the pigs are made a little more accessible by creating fenced-in bottlenecks. "It was a win for the plants, a win for the managers, and a win for the Hawaiian families," Winter said. "They're able to hunt more efficiently."

When I walked out of Kawika Winter's office, I looked at the blue waters around the island. I thought of a line from a Hawaiian chant: *Care for the ocean and the ocean will care for you.* It was time for a swim.

THE SOUTH SHORE of Long Island may seem about as far as one can get from the biogenic beaches and crystal-blue waters of Hawaii. But the coastline here has suffered similar abuse. In the 1970s, sludge was washing up on New York's famous Jones Beach. There were reports of human sewage, oil slicks, medical waste, hypodermic needles, on the beach and in the water, the result of years of sending the city's problems out to sea. After the sanitation revolution of the late nineteenth century, many municipalities began dumping their sewage offshore or piping it into nearby waterways. And in 1938, New York City and New Jersey started shipping their sewage twelve miles offshore to Dumpsite 106, just off the New York Bight. In the 1970s and 1980s, they deposited an average of eight million tons of sludge each year on the continental shelf. (My grandfather, a tugboat captain, might have taken some of the barges out there with his crew.) Much of the sludge stayed at the surface, making bacterial levels rise. The sludge that sank contaminated the seafloor with heavy metals.

The effects of the brilliant, gutsy federal legislation of the seventies—including the Clean Water Act, Marine Mammal Protection Act, Endangered Species Act, and Magnuson-Stevens Fishery Conservation and Management Act—are now visible throughout New York City. The Clean Water Act imposed strict regulations on what sewage-treatment plants and factories could discharge into the harbor. The offshore shipment of sludge stopped in 1992. Water quality improved.

And just as in Kaneohe Bay, the results of new management methods have been astonishing. Menhaden, the forage fish at the base of freshwater and marine food webs, started to return as the rivers cleared up and unsustainable fishing practices were curtailed. After decades of absence, humpback whales returned to the waters near New York City to feed on the large bait balls of these essential

A humpback whale spy-hops in New York City. (Artie Raslich / Gotham Whale)

fish. Fin whales, right whales, and hammerhead sharks are also returning. Common dolphins hunt in the Bronx River; bald eagles and ospreys are again common sights on the Hudson.

There are tens of millions of oysters around New York Harbor, many of them installed by the Billion Oyster Project, which has been working to restore a billion of these classic ecosystem engineers to the city by 2035. We are used to thinking of East Coast estuaries as turbid, rich with phytoplankton. But oyster reefs once filtered entire bays every three days; now it's more like once a year. The project has planted seventy-five million oysters across eighteen restoration sites with the help of high-school students and volunteers. Each oyster can filter fifty gallons of seawater a day, dampen storm waves, and build reefs for other native species such as blue crabs, barnacles, and shore shrimp. Similar efforts are under way in Boston Harbor and other urban areas; people can now swim in water where billions of gallons of sewage a year were once discharged. Whales are back, and so are swimmers and surfers. It's a work in progress, but the coastal economy of these areas has been transformed from one based on extraction (commercial fishing) and ocean dumping (sewage and other pollutants) to a diverse economy that includes stewardship, wildlife watching, and recreation.

Throughout the twentieth century, and even prior to that, every generation was accustomed to loss in size and abundance of fish, birds, and other wildlife. A biologist who started out in the 1960s likely remembered seeing bigger fish and more abundant wildlife at the beginning of his career than at the end. But to his parents, the fish of his youth would have seemed puny and rare. Marine biologist Daniel Pauly called this the "shifting baseline syndrome"—each generation sets a baseline lower than the one before.

But baselines can *lift* too. Around the world, coalitions of government agencies, Indigenous groups, nonprofits, and local volunteers have worked together to restore ecosystems and the animals that build them. A young biologist of the twenty-first century might now see a much greater abundance of wild turkeys, eagles, shorebirds, whales, and seals than her parents or grandparents. There are spiritual and health benefits to having these animals around—as Leroy Little Bear noted, the return of buffalo brought part of the Blackfoot culture back to life. In a sense, people are rewilded. For a new generation, these lifting baselines—humpbacks spy-hopping in the harbor, peregrines swooping over the skyscrapers, alligators swimming up the canal—move nature back to us.

8

The Singing Tree

Conjure the flow of animals in the everyday.

For many of us, whales, bison, and other megafauna are rarely seen, and when they are, they're often viewed from afar, perhaps while we're on vacation. Closer to home, you can see a squirrel leap along the top rail of a chain-link fence. If it's spring, you might see ants: Garden ants moving in a line from a peach rotting in the sink to an invisible hole in the kitchen windowsill. Pavement ants swarming near a sidewalk anthill. Leaf-cutter ants sculpting a hollow in the forest floor.

Numbering in the quadrillions, ants are harvesters, farmers, superorganisms. Some use fungi to convert atmospheric nitrogen into ant biomass, the arthropod answer to Haber-Bosch. Leaf-cutter ants cultivate a fungus garden among the leaves in their nest, and the fungi and bacteria provide food and nitrogen for the leaf-cutters.

By some estimates, entomologist Nate Sanders told me, more than half of the plant species in the understory of forests rely on ants to disperse their seeds. "We once did an experiment where we put seeds on the ground and watched what happened," Sanders said. "Well over ninety percent, and more like ninety-nine percent, of the seeds were carried away by an ant." What does that ant do with the seed? "She carries it back to her nest, rips off the nutrient-rich rewards, and delivers them to her sisters and soon-to-be sisters," Sanders said. Then she takes that seed out of the nest and leaves it in a midden pile along with the refuse they've collected from around the forest floor. "So not only has the mother plant had her seeds dispersed, the ants deposit them in a patch of fertilized ground. Ants truly are ecology's movers and shakers."

For those of us who live in cities and towns, nature's hot spots can seem far away. But there are hot moments too, and for many of us, we can hear them just beyond the screen door. You don't need to go to some far-flung volcanic island or remote national park to see how animals make the planet. It's happening in your backyard, down the road, in the park a few blocks away. As I began writing this book, one of the largest, and perhaps loudest, movements of animal biomass on the planet was set to occur just a few hours from my home. It was in the news just about every day: cicada-mania. Leading up to the emergence in 2021, there was a media buzz that rivaled the mating calls of the cicadas themselves. They headlined morning TV shows, weekly magazines, and newspapers around the United States. One science journalist on PBS argued for empathy for the cicadas that caught the *Massospora* fungus: "Imagine if, after a lifetime underground, you only had a few glorious weeks to live in the sun, eat, and mate. And then your butt fell off."

In May 2021, trillions of periodical cicadas known as Brood X (X for the number 10) emerged from their seventeen-year lockdown, a pulse that occurred from Indiana to New Jersey to Georgia. At that point, I and millions of other humans had been in various forms of lockdown for the past fourteen months. The emergence of the cicadas seemed like an opportunity for me to emerge too—I had to see them for myself.

Cicadas have the longest life cycle of all insects. The adults I was hoping to see had gone underground as wingless nymphs in 2004, before my daughter, now a teenager, was born. They'd spent almost their entire lives beneath a single tree—a sugar maple, perhaps, or a white oak—a foot or two beneath the surface, feeding on the fluids that run through the roots, mostly water and a little bit of sugar and other nutrients. After seventeen years, on the early edge of summer—when the days got longer and a little more humid, and soil temperatures went up to about sixty-four degrees—they would dig themselves out of the ground, leaving behind clay tubes, or "chimneys," that look like simple sandcastles.

In 2004, researchers counted 356 cicadas emerging in a single square meter of land, about the size of a bath towel. In some areas, in a single acre, more than a million of these two-inch insects dug their way up from the soil.

Before I left for Maryland, which has some of the highest densities, I rang up Louie Yang at his lab at the University of California, Davis. Yang had written a *Science* paper on the resource pulses provided by periodical cicadas soon after their last emergence in 2004. We're used to seeing explosive populations of insects, such as spongy moths and locusts, stripping forests of leaves and mowing down croplands, but the effects of mass mortality can be elusive when it comes to insects. "Cicadas have sort of flipped the script," Yang said.

Cicadas eat slowly—chronic herbivory, Yang calls it—over a period of about seventeen years. Very few ecologists look at those incremental changes over time, whether it's the sip of sap from the roots of a tree or the occasional death. Insect outbreaks with intensive herbivory and visible damage get all the attention. "The fertilization effect of most insects is not really noticeable," Yang said, "because dead insects are falling to the ground all the time." It's more like a steady drizzle than a hurricane. Not so with cicadas.

The dramatic pulsed fertilization from dead cicadas makes them different from virtually every other insect on Earth. They are more like salmon, which have a final, single pulse, than like, say, aphids or ants. The emergence of periodical cicadas was documented in the first volume of the world's first scientific journal. In 1665, the *Philosophical Transactions of the Royal Society of London* reported "swarms of strange insects"—that would be twenty-one generations ago for Brood X. Dead cicadas piled up to a depth of three to four inches on beds and dinner tables.

"They would leave an incredible stench," Yang said, but the pile-up got him thinking. "As far as I could tell, people hadn't spent a lot of time studying what happens to them after they hit the ground." Yang found that cicada litterfall, as he called it, increased microbial biomass and the nitrogen availability in forest soils, subsidies that boost growth and reproduction in forest plants. Periodical cicadas thus create strong and reciprocal links between the aboveground and belowground components of the forest. I looked forward to seeing this for myself.

It was my first road trip since the start of the COVID pandemic about fourteen months earlier, which was no time at all for an underground cicada. I got in the car and started a new audiobook, Machado de Assis's nineteenth-century novel *The Posthumous Memoirs of Brás Cubas*. He opens his book with a dedication:

> To the worm that first gnawed at the cold flesh of my cadaver, I dedicate as a fond remembrance these posthumous memoirs.

I was charmed from the start.

DAN GRUNER LIVES in a nondescript brick house in a sea of single-family homes not far from the University of Maryland and the Anacostia River in Silver Spring. I showed up at his house one warm afternoon in early June. Wearing a T-shirt that read MAGICICADA and a baseball cap, he led me out to his back porch to chat. It was as noisy as a crowded bar.

"The street trees went berserk the first week," Gruner half shouted. "Their roots are underneath the road and the pavement of the sidewalks, so it's nice and warm with lots of light." We walked into his backyard. "They're still emerging here in the shade. They waited about two weeks after the main street trees." He offered to show me some cicada chimneys on the path behind his house, but the access was blocked by invasive bamboo. "I don't know how you feel about hopping the fence."

I hopped the fence.

Along the wooded path, the chorus of cicadas seemed as loud as a runway at JFK. Brood X is made up of three *Magicicada* species: "You get the *septendecim* first," Gruner said, "and then you get the *cassini,* and then you get the *septendecula* last." This pattern recurs every seventeen years. As Gruner, an entomologist at the University of Maryland, rattled off their names, they seemed like mere scientific epithets to me; the air was buzzing with cicadas, and they all sounded the same. At first.

Gruner pawed through the leaf litter and unearthed a few chimneys where the cicadas had recently emerged. At first the nymphs are part Joker, part Groucho—red eyes, white body and wings, and thick dark eyebrows. The dark pigment of melanin is expensive, so why waste it underground where it would go unnoticed? Once they emerge, they leave behind brown hollow shells, often left clinging to tree branches as they grow into their new sleek black bodies. Their eyes stay bright red high atop their head in a permanent stare. As I looked around, it occurred to me that these insects were a message from the underground, packaged in an exoskeleton.

Cicadas are clumsy fliers. Just crossing a quiet road can be a challenge for them, never mind a busy highway. I winced when I hit a few on the interstate on the way to Gruner's. They're too slow to avoid predators, and a leading hypothesis for the seventeen-year cycle is that it helps reduce the chances that any of the cicada's many predators could adapt to it. At first, the local omnivores and insectivores responded to the bounty of food. Chickadees, squirrels, spiders, and snakes took advantage of the buffet. A cat snapped up a cicada whole; its whiskers buzzed like a cartoon alarm clock until it yawned and let the cicada go. I saw a grackle swoop in and grab a cicada by the wings. If previous years were any indication, there would be a bird baby boom later in the summer, a result of the extra nutrition.

But once the predators were sated, the rest of the cicadas, part of perhaps the largest insect emergence on the planet, were mostly ignored. Louie Yang estimates that the proportion of cicadas eaten by predators is "vanishingly small, less than a tenth of a percent." So they were left to their own devices. Their long black talons are perfectly designed to hang on to branches and stems, allowing them to remain motionless, expending little energy, for hours at a time.

GRUNER AND I stopped at a silver maple on our way back to his house. It was an old tree, maybe eighty feet tall, leaning up against a rusting chain-link fence. The neighborhood had been ranch-scaped with hundreds, maybe thousands, of one-story red-brick houses. Here in the back of a single-story home in what could have been any American backyard stood a singing tree.

His voice occasionally lost in the wall of sound, Gruner helped me distinguish the cicada calls. One species, *M. septendecula*, sounds like a Japanese laser gun: *Pew, pew, pew!* The sound of *M. cassini* is harder to describe; it's a slow clicking followed by a loud buzz that crescendos as hundreds of other cicadas join in. *M. septendecim* has the eeriest sound, an alien *wheeee-oo* scream in the afternoon. (I've also heard it described as *phaaaaaaar-oah.*) Their sounds overwhelmed every other noise, birdsong and afternoon traffic alike, almost drowning out our conversation. The off-tempo quality of the clusters made it feel three-dimensional. We were in the soundscape of the cicada. Their call-and-response seemed to lift the tree, root and branch, into the afternoon sky.

Gruner checked his phone. "Eighty decibels," he said—as loud as a busy street or a noisy restaurant. We often think of nature as quiet, but a healthy planet is noisy. The air on Surtsey rings with the calls of gulls and fulmars as they bring nutrients in from the sea. A whale exhalation in the fog can be so overwhelming, it envelops you, making it hard to find the direction of the blow or the fecal plume that will eventually follow. The sound of parrotfish crunching corals can fill the reef in Hawaii. At night on the Suwannee River, an owl's call can be so piercing, it causes a shot of terror, then awe. Closer to home for me, in Vermont, frogs announce the seasons: wood frogs quack like ducks at the end of winter as the ground starts to thaw, spring peepers give high-pitched life to vernal pools, and the full-throated trills of gray tree frogs pour in through the bedroom screen on hot summer nights.

Silver Spring wasn't the Maasai Mara or Yellowstone. In some ways, it was more beautiful. The maple we'd encountered vibrated with cicadas, their wings spinning the afternoon sunlight into well-polished bronze. It sometimes seems too obvious to study the most mega of the megafauna, baleen whales, or seek out the most charismatic—lions, tigers, and elephants—when these tiny songsters should command our respect and attention. Awe comes in the strangest of places, and in the most common places too. A green flash over a crumbling seawall; a singing tree in the suburbs. The words *Summer afternoon, summer afternoon* raced through my head. I did not doubt Henry James here; they were "the two most beautiful words in the English language."

The cicadas, of course, ignored us as they went about their brief lives aboveground. These were magical cicadas, with a lot of loud music and a little romance. "It's what they do when they're seventeen years old," entomologist Mike Raupp quipped on Maryland public TV. Making noise boosts their metabolism; the more they sing, the more water—and sugar—they process.

Much as they had been doing underground for years, the cicadas tapped into the trees' xylem—the vessels that transport water and nutrients from the roots to the shoots and leaves. They use the fluid from the xylem, essentially diluted sugar water, for evaporative cooling, much as we sweat and a dog pants. As they sing, they release more water and process more xylem.

"I just got peed on." Gruner laughed. We looked up toward the sun and saw a rainbow of pee refracted in the afternoon light.

Even in dry places like the Mojave Desert, cicadas can access the water table deep underground by tapping into the xylem of mesquite. The more they absorb, the more they release. "It's like an entomological sprinkler system," Gruner said.

To my knowledge, the ecological role of cicada pee remains unexplored. Nonetheless, it would rain down for much of the early summer in Silver Spring.

Soon the females would lay their eggs on pencil-size twigs in tidy white rows of up to six hundred eggs. And when the tiny wingless nymphs hatch, according to Raupp, they'll tumble out, fall eighty feet or so, bounce twice, then go underground for the next seventeen years.

I STOPPED AT a small roadside park after my visit to Gruner's house. Everybody was talking about the cicadas. They had come in like a blizzard or a thunderstorm. There was an enormous walnut tree in the middle of the field, and kids were wandering around it, remarking on the cicadas' calls, their flight, how long they'd be around.

When I got back to my car, I heard a terrible scream. I jumped out of the driver's seat and flailed against my shirt collar. *Wheee-oo!* It buzzed again.

A cicada had crawled into the car. I'm pretty sure my hitchhiker was a *septendecim*, the largest and perhaps loudest of the three species. Gruner was a fine mentor.

Wheee-oo!

As I drove back north, there were reports that cicadas had disappeared in some places. I remembered the deafening call of *Magicicada* on Long Island, just outside of New York City, in 1987. They were almost completely gone in 2020. A friend in Englewood, New

Jersey, on the other side of the city, said that a block party planned in June 2004 had to be canceled because the carcasses were three inches deep and the stench of death was too strong for barbecues. This year, he was sad to report, they were nowhere to be seen, or heard. Populations near Philadelphia and on Long Island failed to emerge.

On the other side of the George Washington Bridge, I heard a sad call from the front of my rental car. My shoulder cicada, or another, seemed to have crawled into the air vent. It buzzed weakly from behind the grille, the sounds getting softer as I entered New York.

MUCH OF THIS book is about long-distance travel, thousands of miles over oceans, up rivers, across continents. But there are local journeys, too, like those made by cicadas, and they matter. Back when I was in grad school at the University of Florida, the mammologist John Eisenberg lectured about sloths. "About once a week, they descend to the forest floor to defecate," he told us. That's a big deal for these leaf-eaters, and a slow—painfully slow, from a human's point of view—hour-long journey from the rich, safe canopy to a latrine in the forest floor. On the ground, tailless sloths leave their feces on top of the leaf litter; those with tails punch holes in the leaves to deposit their poop.

In addition to the usual blood-sucking flies, ticks, and lice, sloths host a large community of arthropods, many of them sloth-specific. Close to a thousand scarab beetles have been found in the fur of a single sloth, clustered around the elbows and knees. Adult moths spend most of their lives in sloth fur, hiding from bird predators and surviving on sloughed skin and the algae that grows on the hair. Many of these insects also spend part of their lives in the sloth dung at the base of trees. Beetles and moths deposit their eggs in the dung; larvae feed exclusively on the feces.

The motivation behind the vertical movement in sloths remains a mystery. Are they marking their territory? It seems unlikely; sloths are arboreal, and a nomadic sloth wouldn't be likely to come across the marking. Others have suggested the opposite motive—by burying their feces, they are trying to remain undetected by predators. Stealth poop. Or were they fertilizing the base of the trees they depended on, sustaining the moths and algae that lived in their fur? Perhaps they ate a little soil on their journey down to Earth.

Every night, trillions of aquatic organisms—from tiny copepods, a million of which could fit in a coffee cup, to sand lance, six-inch-long fish that bury themselves in the seafloor—move from the relative safety of deeper waters, where they can move about unnoticed in the dark, to the surface. Here the herbivorous zooplankton, like copepods and krill, feed on phytoplankton, and the carnivores, forage fish such as sand lance and herring, feed on the zoops. This daily vertical movement of a hundred feet or more is the largest migration on the planet, dwarfing that of blue whales and wildebeests in terms of sheer numbers and biomass. As these creatures move, they defecate and die, like the rest of us, feeding at the surface at night and resting beneath in the deep aphotic zone during the day.

This movement enhances the biological pump and has ramifications for the carbon cycle: the more poop and death near the bottom of this daily migration, the more carbon can be stored or sequestered in the deep sea. Migratory fish can move about one and a half billion tons of carbon each year from the ocean surface to its depths via their poop and migration. (The aviation industry emits about one billion tons of carbon each year.) Not all of this will stay in the deep sea for a hundred years—the gold standard of sequestration—but perhaps the carbon dioxide will stay out of the atmosphere long enough for us humans to get our act together. Perhaps.

"Do I HAVE a wing hanging out of my mouth?"

"It's not a wing." Bun Lai shook his head at the CNN host. "It's a leg."

Brianna Keilar paused to take it out of her teeth. "I don't think my husband will kiss me after this."

"I'll kiss your husband," Bun replied, not missing a beat.

For a hot moment near the center of cicada-mania, Bun found himself the It chef. The insects' predators might have been getting full, but the media outlets were still hungry for cicadas. Bun drove all night to cook cicadas for the *New York Times,* got up early to appear on CNN, spent an afternoon at the studio for *Science Friday* on National Public Radio.

"The idea of eating insects is based on sustainability," he said on CNN. "We're going to have to shift the way we eat"—he removed a cicada leg from the corner of his mouth—"animals."

At the end of my cicada safari in Maryland, I decided to drop in on Bun, one of my favorite insectivores, for a culinary interlude. He had recently closed his family's sushi restaurant in New Haven and started working from home. His new venture, Miya's in the Woods, was a fluid space, more food incubator than farmstead. Colleagues, friends, and interns came and went.

"I am a terrible gardener," he said as we walked out into his yard. We passed a garland of cedar-smoked cicadas hanging on the patio and made a quick left, just a couple of two-legged herbivores grazing his unkempt grounds. Bun, in a black T-shirt and black thick-rimmed glasses, pulled up some bitter cress. We went past some kale casually growing through some wire mesh. "Sometimes I just spill the seeds," he acknowledged.

Bun was more enamored by the lamb's-quarters and dock weed, volunteer plants that had jumped the fence and taken over a patch in the yard. "The best food is not in the garden; it's all around it," he said. Along with the bitter cress and kale, we collected some garlic mustard, mugwort, and mint. "The soup you're going to be having is all weeds."

When we got back to his cluttered kitchen, which also served as an office and a lecture hall, Bun cooked up a quick meal of Faroese salmon and miso soup, rich with the weeds we had harvested. The salad was of foraged greens with a fierce set of bitter secondary compounds.

He poured me a shot of Bun's Flaming Cock sake and handed me a bottle to take home. "That's got autumn olives." The fruit of an invasive tree. We enjoyed the meal, but I wondered about the main course.

"What about the bugs?" I asked.

"I almost forgot the cicadas," Bun said. He heated up a wok on his stove. "Japanese revere insects," he said. He had spent his early years in Kyushu, Japan. "Cicadas and dragonflies represent summertime. We used to climb trees with big nets, catch cicadas, and put them in our bug boxes. We'd just look at them for a while and then let them go. In Japan, they're still coming out in the trillions, even amidst the development." Bun's work—which extends from eating insects to developing a new approach to sushi that emphasizes sustainability and health—takes one small step toward bringing humans into the cycles we've been discussing here. More insects and fewer factory farms and mismanaged fisheries.

Although Bun was an avid proponent of eating insects, he acknowledged that cicadas, emerging every seventeen years, were not going to be a regular item on the menu. He had parboiled the cicadas, then sauteed them with salt and olive oil, inspired by the traditional way of eating insects in Africa. It was important to leave the shells on to retain a crunch. He brought out a cereal bowl filled with pan-fried cicadas, wings still on.

"Are they any good?" I asked.

"They're delicious," he said, handing me the bowl.

We ate them out of the bowl like popcorn. Popcorn with six legs. They were salty, crunchy, a little nutty—delicious in a niche, hot-moment kind of way.

The role of the chef is sometimes to get a leg up on disgust by preparing people for the crunch and the taste. Was the promotion of eating insects working? Bun admitted that the response in the media was too often sensational: "Let's roll 'em in rice!" He wasn't interested in the fear factor. "To me, it's about starting a conversation about eating insects," Bun said. "I am not trying to fetishize cicadas or make them sustainable."

Could cicadas be a gateway bug?

***

"HOT ENOUGH FOR you?"

"It's pea soup out there."

"It's raining cats and dogs."

Everyone notices the weather. Farmers watch it to determine their crop yield. Pilots observe it to plot their course. Just about all of us look out the window—or down at our phones—to decide what to wear.

As we stood beneath the singing tree, Gruner told me that an unusual object had been seen on the weather radar above the forests in central Maryland and Loudon County, Virginia, north of Washington, DC, on a hot and humid day earlier that month. "The weather people think they saw the signal of cicadas," Gruner continued. It wasn't fog or an errant storm cloud—it was buggy.

When animals are abundant, they can be like the weather, moving nutrients and seeds like clouds move water. After the Krakatoa volcano erupted in 1883, fruit bats and birds rebuilt the tropical forests on the island, carrying seeds from figs and more than a hundred other plants across the waters of Indonesia to the barren island. They rained down nutrients in the form of guano too.

As I walked around suburban Maryland, caught up in cicada-mania, I thought, *This is how the world should be, with animals as powerful as a thunderstorm—all-encompassing, seasonal, and not always predictable.* If we traveled back to a time when wild animals were plentiful, not the mere sliver of biomass they are now, they would be a daily force—ecologically, culturally, and socially—just like the weather. Perhaps we could call it meteorzoology.

"The air was literally filled with Pigeons," John James Audubon wrote of a large flock of passenger pigeons he'd seen along the banks of the Ohio River. "The light of noon-day was obscured as by an eclipse; the dung fell in spots, not unlike melting flakes of snow." At the time, three to five billion passenger pigeons, one in every four native birds, flew in the skies of North America. Aldo Leopold described the passenger pigeon as "no mere bird; he was a biological storm. He was the lightning that played between...the fat of the land and the oxygen of the air." It was pigeon weather, with the punch of a hurricane.

Birds still cross the continent, though not in such numbers. (Passenger pigeons were all too easy to hunt and soon disappeared under relentless commercial harvest.) On the day I looked at Birdcast—a website run by the Cornell Lab of Ornithology—82,700 birds migrated over the county where I lived; warblers, flycatchers, thrushes, and tanagers, returning from their winter habitats, flying at about seventeen hundred feet aboveground. It was birdy.

After gray seals returned to Cape Cod in the early 2000s, it got sharky. Great whites moved north in search of the renewed pinniped prey. Just about every conversation at the local bars on Cape Cod turned to the sharks eventually; people grumbled about buying pools for their grandkids, adjusting their times at the beach, losing fish off their lines or out of their nets to the seals. Others embraced the return, with Cape Jaws T-shirts and stranding responses to help seals injured by human activities.

It can be beary in Alaska when brown bears move to the coast, following the salmon runs or searching for human refuse in town. If you're lucky, you might find yourself in "whale soup," a hot spot in a hot moment, with whale blows so thick, they resemble fog.

I KEPT IN touch with Gruner. He told me the skies became "absolute bedlam" soon after I left; his ears rang for hours as the skies filled with cicadas. When the cicadas first emerged, the birds were in all-you-can-eat mode. He'd thought of his yard as a killing ground. Now they hardly noticed the cicadas. The buffet had lost its appeal. But people were still talking about them. My cousin wrote from Annapolis: "Death comes for the cicadas." As it does for us all. Trillions of cicadas fell to the ground and decomposed on the forest floor—an enormous but rare resource pulse. "These interactions are happening all the time," Louie Yang told me, "but they're usually invisible to us."

During the previous cicada emergence, according to Gruner, the media response to the stout, red-eyed bugs had been *Ew! When will this plague end?* But the work of chefs like Bun Lai and entomologists like Mike Raupp and Dan Gruner had helped reframe this moment. Fear and disgust largely gave way to appreciation.

"This year," Gruner said wistfully, "there was a lot of love." Then he got quiet, the silence flooding his office. "Now I have to find a way to fill this hole in my life for the next sixteen years."

9

Cloudy with a Chance of Midges

It was summer in Iceland, the time when midges rose from the lakes, streams, and marshes in the trillions, annoying tourists, long-distance bikers, hikers, and just about everyone except for the resident spiders, insect-eating fish and birds, and a few biologists.

I spoke to one of those scientists a few weeks before my trip to Surtsey. An entomologist at the University of Wisconsin, Claudio Gratton was an assistant professor in 2004 when a colleague knocked on his door to introduce a friend who lived in Iceland. "So, Árni Einarsson comes to my office, opens up his laptop, and says: 'I work at this lake in northern Iceland called Myvatn. In Icelandic, *my* means "midges" and *vatn* means "lake." It's famous for these massive insect emergences.'"

That caught Gratton's attention. Einarsson had been studying midge emergence and its effect on predators. Gratton asked about the movement of insect biomass out of the lake and onto the surrounding land. "And he kind of looked at me in his very Icelandic way and said, 'No, I don't have any idea what goes on. It becomes less interesting when they get into the grasses.'"

To Gratton, that was just when the excitement began. "It's probably one of those things that as a junior faculty, you should never try to do," he said, "start a brand-new project on a pretty speculative idea in a foreign country." But the midge emergence sounded like it could have broad ecological consequences. Gratton had a few dollars left over from his start-up fund and decided to risk them on a trip to Iceland.

Most of the animals on Earth, by both numbers and weight, are arthropods—insects, crustaceans, and their relatives. Ecosystems around the world, from mountain streams in Ecuador to lakes on the edge of the Arctic, depend on insects to break down and disperse nutrients. Many of these arthropods are not much bigger than punctuation marks—some the size of a comma, others the size of an exclamation point.

!

A single midge emerging from Myvatn after several months of feeding as a larva at the lake bottom wasn't going to make a world of difference to the landscape or the predator community. "But there are just *so* many midges that there's no way that all the predators out there could eat enough of them," Gratton said. The vast majority die on land around the lake. "When they die, they decompose." He wanted to know what happened to the midge-derived nutrients: How did they influence the plant communities, microbes, and decomposers? "That's where the magic starts to happen," Gratton said.

Gratton first went to Myvatn in 2006. It was a good year for *Tanytarsus gracilentus,* the dominant midge among the thirty or so species found in the lake. In pictures, Gratton is joyously walking through a thick buggy fog with his sweep net. Einarsson posted a selfie taken during the long summer magic hour: a Chuck Close portrait in midges.

"Being inside one of these swarms, hearing them, smelling them, and just watching them undulate in the grass was like nothing I've ever seen before," Gratton told me. He returned for a few weeks every summer for the next twelve years. "For me, it's like going to an art museum for an art lover. It's my Met or my Louvre."

IF I COULD get to a place as inaccessible as Surtsey, I thought, it would be relatively easy for me to visit this museum of imaginative midge leaps. But from the start, it was riddled with challenges. I called around but couldn't find a rental car, and COVID was spiking in Iceland. Everyone was on vacation or holed up in their country homes, and Árni Einarsson was proving difficult to track down, though he had warned me that this would be a low-midge year. Even the local research center was under lockdown. For a small fortune, I managed to lease the last rental car in Reykjavík online, with a hatchback big enough to sleep in. Road trip.

At Surtsey, Bjarni Sigurdsson described Myvatn with mime-like motions indicating that moving through the midge swarms was like passing through a series of thick velvet curtains. But there were no dipteran draperies when I arrived. Einarsson's forecast for a low-midge year had been correct. I drove around the lake and hiked along the paths lined with the diminutive Icelandic birch—if you get lost in a birch forest here, the saying goes, stand up—and the occasional buttercup in bloom. There were a few bugs flying close to the ground when I knelt out of the breeze, but no aggregations.

I took another turn around the lake. Myvatn was formed by a volcanic eruption about twenty-three hundred years ago, and there were sharp basaltic columns around the edge of the lake and vast lava fields spreading into the distance. The lake is about five miles long and very shallow, fifteen feet deep or so. I pulled over to get some gas. There were a few ravens picking through the trash and, for a moment, a swirl of flies around the station. But there wasn't a midge in sight. Peter Matthiessen's *Snow Leopard* came to mind—an entire book in which the author doesn't see the eponymous beast. But there were only a couple thousand of those wildcats on the planet. Midges numbered in the trillions in Iceland, and yet I was mostly seeing rain.

Myvatn *is* a midge hot spot—but I was there at a cold moment.

Bjarni Sigurdsson e-mailed me that evening. "You're probably the only person in Iceland who's sad that there's not a lot of midges."

IN A GOOD year, larval midges spend the spring scraping algae and detritus off the lake bottom. As larvae, they spin small silken shelters that they reside within, crawling out occasionally to feed. At first, they generally eat dead bits of organic matter, but as they get bigger, they start feeding on algae as well. "Living inside of these shelters, they poop and excrete," Gratton said, "and these tiny little tents become a place where algae can grow quite rapidly." The larvae create a nutrient-rich habitat that alters the lake bottom. It sounded a bit like midge gardening or choreography, much as we've seen with corals, parrotfish, bison, and whales. The midges reach out of their shelter, scrape off the algae that's growing on the outside, and eat it. The larvae poop, molt, and go through a series of stages, getting bigger, eating more.

At this time of year, the larvae dominate the ecology of the lake, with midges making up about two-thirds of herbivore biomass in Myvatn. Fish and birds depend on them for food. Then in late May, the adults break the surface and take flight. "As soon as they hit the surface, the shell splits open, and the adult midge will fly out of there as quickly as possible," Gratton said. "They do this very rapidly because the longer they sit on the water, the more likely they are to get churned up by the waves or picked off by ducks or other birds." The terns sweep in and snatch them from the surface.

The numbers start to build, perhaps slowly at first. Each individual, made up of benthic algae, takes to the air, an exclamation point with tufts of antennae for a head.

!

!!

!!!

The adults move out over the surrounding heath and grasslands. The males form mating swarms, probably because they're more visible or attractive. *Tanytarsus,* the dominant midge of Midge Lake, forms a ground-covering fog that looks like a wool blanket tossed over the landscape, with the males pilling up over the surface.

"They're not very strong fliers," Gratton noted. When the wind dies down, they start making aggregations. "It reminds me of the smoke monster in *Lost* coming up out of the grass and moving around as the wind shifts. It's really eerie and very comforting. As an entomologist, this is the place to be."

The other common species, *Chironomus islandicus,* forms mating pillars up to ten or fifteen feet in the air. The pillars swirl around "like devil's tongues during a wildfire, spiraling up toward the sky," Gratton said. "As soon as the wind blows, the whole pillar collapses, and they all kind of settle back into the grass until it calms down again. When they start to come out again, you can hear them droning. From a distance, it's like someone's got a lawn mower going, but you don't quite know where it is."

Sure, there are plenty of bugs, but, as one ecologist asked, are there enough to raise an ecological eyebrow? Gratton and colleagues looked at changes in the landscape and the animals surrounding the lake associated with midge emergence. The spiders ate fewer leafhoppers, their usual prey, in outdoor experiments Gratton's team designed. He noted that it didn't matter whether he added a hundred midges or ten; just their presence was enough to distract the spiders. They spent less time searching for their normal prey. They saw

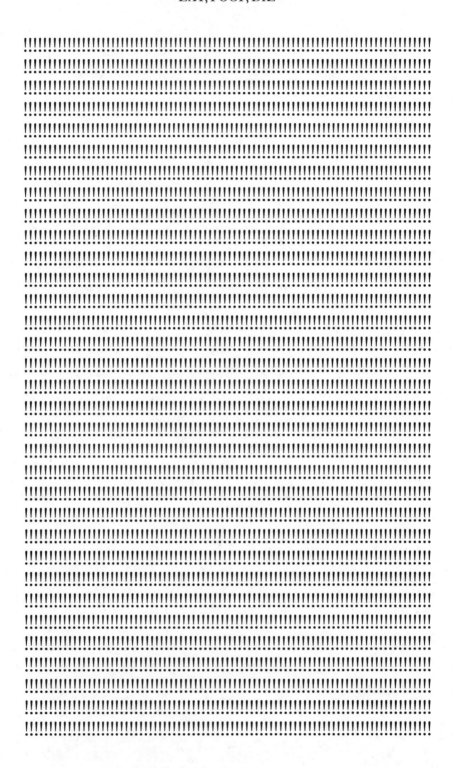

the same pattern in the grasses around Myvatn. "Spiders seem to get hypnotized by the sound of the midges' beating wings," Gratton said. Once the spiders heard it, they waited for their midge meal, ignoring the resident terrestrial insects. As a result, springtail, mite, and aphid populations increased.

On Surtsey, you can tell where the birds are by the spring in your step on the grasslands. You can feel their absence as the lava crumbles under your feet. The midges of Myvatn similarly transform their environment. The tiny lake-grown midges add up. As the carcasses start to decay, you can smell it; it's like a can of rotting tuna, as Gratton described it. In a high-midge year, Gratton estimated that more than a hundred pounds of midges rain down on each acre, in a band that stretches along the shore for about a hundred yards from the lake. About two hundred yards from the lake's edge, it's more like a drizzle. Few midges make it farther than that, and the difference is stark as the grasslands give way to nutrient-poor heathlands. The area beyond the midges' reach feels barren and unproductive. No matter the season, it's clear where these tiny insects swarm: grasses thrive in areas with them and wither in areas without.

"I think of the landscape patches as a tapestry," Gratton said. "The midges bring the essence of the lake onto the land. You could grab a spider and find that it's made up of nutrients from the lake. The spiders are woven from material that is not where they come from. Myvatn has taught me that we're not as discrete as we thought."

During peak midge season, more than a hundred and twenty tons of the tiny insects fall around the lake. That's the weight of more than five hundred thousand Big Macs or two million meatballs rolling around the lakeshore, depending on your recipe.

Gratton and his colleagues estimated that these midges deposit about eleven pounds of nitrogen per acre per year, about the same as the seals on Surtsey but not as much as cicadas in Kansas, at thirty pounds per acre. Salmon carcasses deliver seventy pounds per acre.

Gratton was excited by the findings, and since Myvatn was sur-rounded by private land, including lots of farms, he thought he should talk to the farmers and tell them about what the midges were doing. It turned out they already knew. "We would sit around the kitchen table," Gratton said, "and we'd asked them if they knew anything about these midges. And through broken English and Icelandic translation, they said, 'Oh yeah, we have this historical knowledge that goes back many generations from family to family. If midges are flying in late May and early June and then it rains, that's when we get the biggest production of grass. We can harvest much more hay in those years than we do in years without the midges. We call it *mygras,* or midge grass,'" they told Gratton. Those were good years, they'd said. They could feed more hay to their sheep.

After he heard this story, Gratton decided to test the impacts of these carcasses by scattering some frozen midges in a low-nutrient heathland, where the vegetation was sparse, hardly growing up to his ankles. After two years, the midge effect was obvious. "You could clearly see that the areas that had gotten more midges had much more grass than the areas without midges," Gratton said. "You didn't need statistics." (Though of course they did the measurements and analyzed the data.) By year four, the grass was almost up to their knees. The nitrogen, phosphorus, and carbon in the bodies was stim-ulating the productivity in plants.

When he went back and gave a presentation to the farmers, they said, "That's what we thought." They were happy to see the science back up what they had observed for generations. It isn't surprising that farmers often pick up on these patterns before scientists do. They need to. In the days, not so long ago, when Iceland was isolated and relatively poor, "if you didn't put away enough hay in the winter for your animals, you could lose some," Gratton said. You couldn't go to the market and buy more hay. "In some ways, midges contributed to the survivorship and well-being of people through the harsh winter months," Gratton noted. "They were like a lifeline for people."

Examples of Nitrogen Movement, or Resource Subsidies, by Animals, Including Human Agriculture

Habitat/Location	Movement	Source	Pounds of Nitrogen per Acre
Surtsey grasslands	Ocean to island	Seabirds	60
Surtsey shoreline	Ocean to island	Seals	12
Florida beach	Ocean to dune	Loggerhead sea turtles	27
Alaskan forests	Ocean to river and forest	Salmon	70
Maasai Mara	Grasslands to river	Hippopotamuses	624
Maasai Mara	Grasslands to river	Wildebeests	294
Conventional cropland	Human agriculture	Industrial fertilizer, chicken or cow manure	100
Permanent grasslands	Human agriculture	Industrial fertilizer, chicken or cow manure	25–50
Walking trails	Human subsidies	Pet dogs	110
Kansas	Underground to aboveground	Brood IV cicadas	6
Myvatn	Lake to land	Midges	11

Sources include rough estimates for croplands and grasslands from Einarsson et al., 2021, Subalusky et al., 2019, and references in earlier chapters for wild animals. Original measurements for hippos and wildebeests are in grams per square meter, so comparisons are a little loose.

On my last night in Myvatn, I parked at the lake edge. I was surprised to hear the wail of a loon, a call I associate with the long evenings of the Adirondacks. I watched the loon swim by with its pen-like beak. An arctic skua flew overhead. Gossamer stretched over the vegetation and across the road. Then it happened: A column of midges formed, a micro-swarm, backlit by the nine p.m. sun. The shrubs were aglow, and the swarms seemed to breathe in the gentle breeze, coming together and falling apart. The insects landed on the windshield like a soft drizzle.

If the Louvre was closed, I could at least visit the National Midge Gallery of Iceland.

There were two types of midges crawling along the car window, one barrel-chested, short, and squat; the other long and leggy, like the King's Guard at Buckingham Palace, but with a big head of translucent antennae in place of the bearskin hat. I watched them race up the window, then launch into a pointillist landscape.

Out of this microburst, I tried to conjure up a hurricane. As I got out of the car, the flies started to cluster around my eyes and crawl into my ears. Isolated swirls formed a curtain as the wind died down. There was a swarm, perhaps in the hundreds of thousands, as I looked toward the sun, lingering on the horizon.

My heart sank when I stopped at the well-appointed Myvatn Research Station the next day. There were about thirty midge species found around the lake in a typical year, I was told, but *Tanytarsus*—the species I was there to see—was strikingly low that summer. The emergence of midges in Myvatn is dynamic, depending on how much food is available in the lake; it doesn't occur in predictable cycles like, say, the seventeen years of the Brood X cicadas. It was unclear if the new lows were part of a disturbing trend or a normal reflection of the year-to-year variation, as Myvatn's farmers have observed for centuries.

The squat and shiny insects I had seen were blackflies, biting midges—tiny hyenas running across the savanna of the car window. "That's classic blackfly behavior," Gratton said when we chatted later. I had probably been watching the males, which were mere nuisances. "They get in your ears and your eyes," Gratton said, "but they don't bite." Only females need a blood meal, taking a little chunk out of your skin, to breed. These were not the nonbiting midges I had come to see—I could find plenty of blackflies back home. "They give a little pop when you crush them," Gratton said. "And it just makes a giant mess."

On the bright side, the swarms I had seen were likely *Chironomus islandicus*. During boom times, most of the biomass would be *Tanytarsus gracilentus*, the *My*s of Myvatn, Gratton's *Mona Lisa*. But in this off year, I had been lucky to observe these relatively small six-foot columns, no matter how ephemeral the swarms.

I saw a few clouds of midges as I left the lake. Probably more blackflies. *Tanytarsus*, one of the most abundant insect species in Iceland, had eluded me. If I were a nineteenth-century sheep farmer, I might have starved.

Os Schmitz poked around the edges of an old field in Burlington, Vermont, with a large white sweep net in his hands. Hubbard Park, butting up against housing developments and woodlands, isn't as exotic as Myvatn or Yellowstone, but it has all the drama if you know where to look. The battle was being fought around shoots of Kentucky bluegrass and long green stems of goldenrod.

Schmitz and his lab mates were searching for a sit-and-wait predator, *Pisaurina mira*. *Mira*, as they called it, is a long-legged, sleek-bodied nursery web spider with a dark brown racing stripe running the length of its cephalothorax (head and abdomen).

One summer afternoon, we had swept the area around this particular study site, a suburban field with a view of Lake Champlain and the Adirondacks, but come up empty. "Once you mow," Schmitz said, "you lose the structure, and there's nowhere to hide. That's the problem with exurban lawns; people keep them mowed, so they lose the insect diversity." So we looked through abandoned lots, cemeteries, and a pocket park under the flight path of Burlington International Airport. Found on the grasses, high weeds, and low shrubs of eastern North America, *Pisaurina mira* was quiet that day, playing hard to get.

One of Schmitz's postdocs scrolled through the website iNaturalist, looking for photos and locations. "There's something lottery-esque about this," Schmitz said as we swept the fields in search of spiders. A professor at Yale, he wore a wide-brimmed hat and a light blue long-sleeved shirt. There was something charming too. An initiative of the California Academy of Sciences and *National Geographic,* iNaturalist hosts ninety million images tagged with locations and dates. It's an app used by schoolkids, amateur naturalists, and, judging by where we drove that day in the team's white van, world-renowned ecologists. We would find a few of the nursery web spiders in our travels that afternoon.

Schmitz has written widely on elephants of the African savannas, howler monkeys in the Amazon, and musk oxen on the Arctic tundra. But these days he's mostly known for the landscape of fear he has created in the unkempt fields of New England—arachnophobia among the grasshopper set. Compared with studies on animal ecology set in seemingly inaccessible landscapes such as Surtsey and Bristol Bay, Schmitz's work on common grasshoppers, spiders, and plants that many would dismiss as weeds might seem positively pedestrian. But there are advantages to creating these miniature ecosystems where animals can easily be added and removed. Top-down controls and trophic cascades, it turns out, are just as important on the edges of suburban lawns as they are in areas like Yellowstone and the Maasai Mara.

After several days of searching, Schmitz and his colleagues found enough spiders to create a *Pisaurina* paradise—several mesocosms of chicken wire and thin mesh, about three feet high, held down with rusting black binder clips. They stood out in the field like miniature Icelandic basalt columns—but here they surrounded Kentucky bluegrass, bedstraw, and goldenrod, which were coming into full bloom when I visited. The enclosures were designed to hold red-legged grasshoppers, a short-horned generalist, and two apex carnivores: *Pisaurina mira*, sit-and-wait predators that hang out all day, like the cougars of the arachnid world, and *Phidippus clarus*, a wandering jumping spider.

The jumping spiders had an impact on grasshoppers in the familiar way: they killed and ate the plant-eaters, so the plants got a break. Ecologists describe this as top-down forcing; the influence that predators have on their prey can affect the entire food web through consumptive interactions, as we saw with wolves. These trophic cascades are widespread, and we'll discuss one of the most famous interactions in the ocean soon, but these interactions can be nuanced too.

Things were different for the nursery web spiders. At first, Schmitz was surprised that the grasshoppers were consuming the goldenrods that gave them cover from the predators that hunted them. *Well, that's stupid,* he'd thought. *Why are they eating their refuge?* Then he realized: It doesn't matter if you have six legs or two—fear increases metabolism. *Pisaurina mira* is quiet and still but also deadly. Under the steady eight-eyed gaze of the sit-and-wait predators, the grasshoppers started stress-eating the carbon-rich goldenrod, herbs that filled their guts faster. "Whenever we're stressed," Schmitz said, "we crave carbs"—regardless of the consequences.

Schmitz had created a landscape of fear, a "nonconsumptive interaction," as he calls it, where predation changes prey behavior and physiology. Even spiders with their mouthparts glued together and no ability to kill—Schmitz calls these risk spiders—had a huge effect on the grasshoppers, changing their behavior throughout the day, throughout their lives. As the grasshoppers ate more goldenrods—with more carbohydrates and less digestible nitrogen—their chemistry changed. Their poop changed, and their carcasses changed.

To see if these small changes could add up to a big impact, Schmitz and his colleagues buried some stressed-out grasshoppers beside those who'd had a relatively chill, predator-free life. The bodies of the stressed-out grasshoppers had more carbs, and thus more carbon, and fewer nutrients, like nitrogen. It turns out that the landscape of fear persisted in the ecosystem. In "Fear of Predation Slows Plant-Litter Decomposition," he showed that changes in the carbon-nitrogen ratio changed the grasshoppers and the grasses. Predators played a role in the carbon cycle even without consuming their prey.

The soil around the stressed grasshopper carcasses had a lower nitrogen content than soil in areas where the more relaxed insects were buried. The lack of nutrients had a profound effect on the ecology of the field, slowing succession—the changes in plant communities—to a crawl. With spiders and their grasshopper prey, it could take fifteen or twenty years for a forest to take root. Without the spiders, it might take only six or seven years for an old field to fill in with an early successional forest.

There was a bigger pattern at play here too. There was about a 40 percent increase in carbon retention in plant biomass when the spiders were present, largely because there was more carbon stored in grass and belowground root systems. These changes in carbon uptake, allocation, and retention were largely driven by fear. "These tiny little creatures that you don't really see unless you walk carefully through the vegetation," Schmitz said, "have a dramatic effect on ecosystems."

Schmitz and his colleagues later coined the term *zoogeochemistry*, the concept that is the beating heart of this book. "The prevailing thinking in biogeochemistry is that it's plant-soil microbe interactions, because those are the dominant players in the field," Schmitz noted. But the animals are playing a critical role too. With sweeping multiplier effects, animals, from whales to spiders, have a profound effect on the biogeochemistry of plants, soils, and oceans. "That's why we emphasize this concept of zoogeochemistry," Schmitz said, "to get the geochemists to pay attention to animals."

It certainly got the attention of ecologists. As classic sit-and-wait predators, mountain lions interact with other species in hundreds of ways. By killing elk and other prey in their preferred hunting areas, these large cats create nutrient-rich hot spots of nitrogen and carbon, released through the carcasses of their prey and through the poop and pee of the cats and local scavengers. These hunting grounds, or "kill gardens," end up lush with nitrogen-rich plants, which in turn attract more herbivores, which then eat, poop, and possibly die in the area, continuing the cycle. Mountain lions act like farmers, creating hot spots that attract more elk and deer. Larger predators influence the landscape too.

ANIMALS ARE MORE than numbers and vectors. Their personality traits, preferences, and traditions have different impacts on ecosystems. This is true for chimps and dolphins as well as for grasshoppers and crayfish. During my chat in the Maasai Mara with Amanda Subalusky, she told me that a colleague's experiment showed that crayfish personalities influenced ecosystem function. Aggressive crayfish had turbid tanks because they kicked up the sediment. Passive crayfish of the same species had clear tanks. "That messes with your head," she said. Chris Dutton, her spouse, continued, "In camp, we have this little dwarf mongoose that has lived there for a decade. We know it's the same one because he is missing a tail, probably from scrapping with somebody else. He plays a big role in camp, because he picks up any trash that someone drops immediately—that keeps the ants away for the most part...But how do you model that one little important player that has a big influence?"

One way is to break down personality into a few traits, such as boldness, aggressiveness, and (my favorite) exploratory tendency. Individual animals, from aggressive mice to pioneering seabirds, shape ecosystems through traditions or by taking risks. Some wolves, mountain lions, and badgers have a taste for beavers, which gives them an outsize effect when they kill these ecosystem engineers. A bold mouse might hoist a big acorn far from the mother oak to stash it away, but the journey can be risky, filled with predators. A shy squirrel might survive longer, hiding nuts closer to home. One of the only ways that tree populations can move is through animals. A forest with bold and shy rodents is a forest that can adapt to climate disruption. Its future could be in the grasp of these seed-dispersing rodents.

States of emotion, or something like them, have been observed in lots of animals, and they can change with experience. Shaking a honeybee can induce a kind of pessimism; the honeybee assumes the worst of new experiences. If bumblebees are given an unexpected reward, they become more optimistic. "Good moods are also seen in fish," philosopher Peter Godfrey-Smith noted in *Metazoa*.

When we look at a trillion animals migrating up and down in the ocean, they might seem to merge into a single entity. But each individual is motivated by its own levels of hunger and tolerance of risk, just as we observe differences in boldness and appetite among our pets and our friends and family members. Not every animal moves. "The fear factor plays out in daily vertical migration," Schmitz told me, the greatest movement of animals on the planet, from the ocean surface by night to the safety of the deep dark by day. Some individuals skip out on the daily commute entirely, perhaps because they're not hungry enough to make the journey so it isn't worth the risk of getting eaten alive at the ocean surface.

This fear factor affects humans too. When you see yourself as meat, things change. When you swim with sharks or hike in the presence of brown bears, your awareness sharpens. *This is not really happening,* Australian ecofeminist and philosopher Val Plumwood thought after being struck by a crocodile in the Northern Territory. "Few of those who have experienced the crocodile's death roll have lived to describe it," she later wrote. "It is, essentially, an experience beyond words, one of total terror. The crocodile's breathing and heart metabolism are not suited to prolonged struggle, so the roll is an intense burst of power designed to overcome the victim's resistance quickly." Plumwood not only survived but came up with new ideas about death and the ecological order. "I glimpsed beyond my own realm a shockingly indifferent world of necessity in which I had no more significance than any other edible being."

"Welcome to my world," said the grasshopper to the philosopher.

Schmitz noticed that individual grasshoppers in his studies had different reactions to spiders. Just like seabirds, grizzlies, whales, and scientists, grasshoppers have distinct personalities: some are bold, others timid; some are explorers, others loners. When a spider is nearby, some go about their business casually, ignoring death. Some cower and rarely stray from their cover. Many stay cautious whether the spider is in sight or not.

"PTSD is adaptive if you're in danger," Schmitz said, "and a predator is always around."

"Hey, bear!"

10

The Otter and the H-Bomb

E ven before the hatch was shut on the Lockheed C-130 Hercules, the cargo started to stink. Designed as an airborne emer-
gency room, the transport's hold had space for almost a hundred
medevac litters. Now it carried fifty-two sea otters that had been airlifted from Alaska's Aleutian kelp forests, all lined up in bathtub-size
kennels. Their lush brown pelts—"soft gold," once the most valuable
fur on Earth—were slick with feces.

The plane was headed east to Sitka, Alaska. Though the city was
built on the pelts of sea otters in the late eighteenth and early nineteenth centuries, not a single otter had been seen in Sitka Sound
in fifty years. In fact, otters had been hunted out of thousands of
miles of coastline, from southeast Alaska to British Columbia, Oregon, and Washington. Ninety-nine of every hundred otters had
been killed. By 1911, when an international treaty was signed to halt
commercial harvest, only a few hundred otters survived in isolated
populations in California, Alaska, Russia, and Japan. In the late
1960s, the U.S. Atomic Energy Commission, preparing to detonate

the largest underground nuclear explosion in American history, was about to change that.

MORE THAN TWO hundred years before the otter airlift, Peter the Great had dreamed of sending a Russian expedition to explore the area between Asia and North America. He had watched other European powers colonize the Americas, which he believed were connected to Russia's eastern shores. In 1724, close to death, the czar commissioned Danish commander Vitus Bering to lead the first Kamchatka Expedition to the Americas. It failed. The land bridge that Bering was hoping to follow to North America had been submerged beneath cold North Pacific waters at the end of the Pleistocene, about ten thousand years earlier.

Bering tried again in 1739. On this expedition, Georg Steller, a young German naturalist appointed to join the mission, described several species that were new to Western science, including a sea eagle, eider, and jay that would eventually bear his name. During some downtime in Kamchatka, he provided the scientific names of five abundant North Pacific salmon: *Oncorhynchus tshawytscha* (Chinook), *O. keta* (chum), *O. gorbuscha* (pink), *O. kisutch* (coho), and *O. nerka* (our beloved sockeye).

The expedition, the first European ship to reach continental Alaska, also reported on the many marine mammals in the region, including whales, seals, and what would eventually be called Steller's sea cow, a thirty-foot relative of the manatee that was as wide around as it was long and so fat that it couldn't dive. Steller was part naturalist, part salesman—not an unusual combination in the eighteenth century. For Russian investors, his greatest find was the abundant sea otters in the region. "These animals are very beautiful," he wrote, "and because of their beauty they are very valuable." The hairs were "very soft, very thickly set, jet black and glossy." Unlike other marine mammals, sea otters don't have a thick layer of blubber; their dense fur, with about 850,000 hairs per square inch, provides warmth and buoyancy. That's about eight times denser than a typical human head of hair. I was given a sea otter pelt to hold in Alaska, and I have never felt anything like it, so luxurious that I can still recall the touch on my fingertips.

Bering died on the return trip and was buried along with other sailors in shallow graves soon picked over by arctic foxes. The vast area he traversed would later be named for him: Beringia. By the 1760s, a couple of decades after the expedition, the beautiful sea otter furs reached the aristocracy in Canton, China; the nobles fell in love with them. A sea otter fur became as valuable as ten beaver pelts, the pillar of the fur trade in North America at the time. Hundreds of thousands of sea otters were killed and sold by Russian companies at enormous profit. Many of the pelts passed through Sitka, in southeast Alaska, home of the Tlingit and occupied by Russia. Over time, the American presence grew in Alaska; small vessels left New England and sojourned in the Pacific Northwest, trading for skins with the Tlingit, Nootka, and Haida. The ships then sailed to Canton (now Guangzhou) to offload the pelts.

More than a million sea otters were killed by European and Native hunters during the height of the fur trade. By the 1840s, they had been eradicated from most of their range, and after years of warfare and disease among the Indigenous people in the Pacific Northwest, the number of hunters had declined. With fewer pelts, Russia saw little value in its overseas territory. American hunters were still harvesting the last otters, so the United States made Russia an offer for the land. According to Jan Straley, a marine mammologist in Alaska, "Sea otters are the reason we were not sold to Canada."

"No ONE ON planet Earth could speak truth to power like John Vania," Jerry Deppa told me when we met at the Backdoor Café, tucked behind a bookstore in Sitka. Deppa worked for Vania as a field biologist at large with the State of Alaska in the 1960s and 1970s. "The Atomic Energy Commission was getting kicked out of

Nevada by a combination of politicians and organized crime, not an unusual thing," Deppa said. One botanist quipped that the search for a new remote nuclear test site was prompted "partly to prevent jiggling of the gambling tables at Las Vegas." America's hydrogen bombs had gotten so large that they couldn't be safely detonated on the continent.

"They thought nobody would make a fuss in Alaska," Deppa told me. After Lyndon Johnson's advisers assured him that there would be no public relations problems, he authorized a new package of nuclear tests to begin in 1965. At first, it seemed that his advisers had been right. "The governor: No objection," Deppa said. "Congressional delegation: No objection. Commissioner of Fish and Game: No objection. Nobody dared to stand up to the Atomic Energy Commission and say, 'You're not welcome here.' They rolled over like a bunch of trained pups."

Alaskan state biologist John Vania invited himself to a meeting in Anchorage between the AEC and Alaska state representatives, according to Deppa. The commission discussed their plans to test three bombs, each one larger than the last, on Amchitka Island, on the western edge of the Aleutian Islands. Vania told the representatives from the AEC about an impact from the test that they hadn't even considered: the adorable tool-using sea otters of Amchitka, celebrated in a recent TV documentary. "Any otters that find themselves submerged when the pressure wave comes through will hemorrhage and die an unpleasant death," he told the commission. "This will be witnessed by a lot of folks. So that's the bad news. The good news is we have a place that needs these otters, and we have the wherewithal to move them. We will evacuate the otters for you. We just need the financing."

At the time, Amchitka had the largest population of sea otters on the planet. U.S. Fish and Wildlife Service biologist Karl Kenyon, who knew more about otters than almost anyone, had pushed to transport Amchitka otters to southeast Alaska to help reverse the disastrous legacy of the maritime fur trade. A few attempts had been made, but they had been sporadic, underfunded, and largely unsuc-cessful. (This was before oil was discovered in Alaska, and the state was still relatively poor and undeveloped.)

The AEC had begun working on Amchitka even before Vania had his first sit-down. The island had a deep harbor, a World War II runway, and plenty of Quonset huts. (It had been used as a U.S. Army airfield to attack Japanese forces in the Aleutians.) An initial explosion in 1965 set the stage for the Cannikin detonation, sched-uled for 1971, which would be one of the world's largest. Before this blast, the AEC ordered a thorough survey of the geological and nat-ural history of the island. A community of scientists gathered to-gether, a remote Manhattan Project of about eight hundred people, including oceanographers, limnologists, botanists, ornithologists, and ichthyologists. There were principal investigators, students, and support staff. "We were a pretty close-knit community," ecologist Jim Estes, a graduate student at the time, recalled.

But outside of Amchitka, things were changing. Vania's words would later prove prescient. Widespread protests against the second and third nuclear tests caught the AEC and federal government off guard. An upstart environmental organization based in Vancouver, then known as the Don't Make a Wave Committee, also had an eye on the island. The young activists were planning to stop the detona-tion by confronting the U.S. government from an eighty-foot halibut seiner they renamed *Greenpeace* for the voyage.

Thousands of bloody otter corpses would not help the Atomic En-ergy Commission's cause. The governor of Alaska, Walter Hickel, wrote to the AEC echoing Vania's request that the Amchitka otters be transplanted to other areas in Alaska before testing, in part to save their valuable furs.

Alaska's small Fur and Feathers Division soon found itself with a blank check from one of the most powerful agencies in the world. "It was El Dorado for the sea otter program," Vania's friend and col-league Skip Wallen told me. "To move the otters from Amchitka to southeast Alaska, we needed an airplane." The AEC immediately offered to supply them with a C-130 Hercules, a plane that could move more than fifty otters in a single flight. It would have taken an entire field season of flights in smaller planes.

Though no one knew it at the time, the sea otter airlift would represent one of the first—and most successful—cases of rewilding on the planet. The return of these predators would show us just what animals could do in restoring wild seascapes.

IN THE SUMMER of 1968, with funding secured, John Vania and his colleagues set gill nets—monofilament netting typically used to trap fish—above Amchitka's kelp forests. Otters spend a long time grooming at the surface of the water, and when they hit the nets, they often respond by twirling their bodies, ensnaring themselves even further. Once they were hauled out of the water, the otters were cut out of the lines and held in two large above-ground swimming pools awaiting evacuation in the cargo deck of the Hercules plane for points east.

After the sea otters arrived in Sitka, Deppa moved them to the smaller Grumman Goose—basically a boat with wings. The amphibious plane was a practical form of transport in Alaska, which had few long-distance roads. When the Goose landed on the glassy waters of Sitka Sound, Deppa had the privilege of opening the kennel doors. More than forty years later, wearing a green Australian bush hat, thick black glasses, and leaning on a walking stick he had cut from a red alder, Deppa recounted the story as if it had just happened: "They had just gone through hell," he said. "They had come close to dying in an atomic apocalypse." The otters were stressed, their once beautiful coats covered in feces. Before the release, some scientists were concerned that they might try to swim back to their native Aleutians.

Once they hit the water, Deppa doubted they would search for home. "Their eyes are as big as saucers," he said, "because in the process of cleaning their fur, they look down at a carpet of huge red and purple sea urchins. And the next time you see them, just out past the wingtip, they have a big sea urchin on their chest. They'd just come to paradise."

Between 1965 and 1972, the Fur and Feathers Division translocated 710 otters from Amchitka and neighboring islands to southeast Alaska, Washington, and Oregon. Forty-three Amchitka otters were released on Vancouver Island with the support of the Fisheries Research Board of Canada and the British Columbia Fish and Wildlife Department. In the decades that followed, these otters and their offspring have spread hundreds of miles from the original drop-off points.

Greg Streveler, a scientist at Glacier Bay National Park, was on a Grumman Goose that dropped twenty-five otters in Dicks Arm, a bay north of Sitka, in 1968. "Nobody in those days," Streveler said, "had the slightest inkling of the revolution they were going to bring with them."

ON NOVEMBER 6, 1971, after five years of preparation and the evacuation of a large number of otters, the Cannikin warhead was detonated a mile below the surface of Amchitka. The largest underground nuclear explosion conducted by the United States, the blast was 250 times more powerful than the bomb dropped over the city of Hiroshima. It measured 7.0 on the Richter scale, blowing open a crater more than a mile wide and sixty feet deep.

James Schlesinger, the chairman of the AEC, brought his wife and two daughters to the island to highlight the safety of the test. In a blockhouse twenty-three miles from the blast site, his nine-year-old daughter said the shock waves felt "like a train ride."

The denizens of the surrounding waters weren't so secure. Nearly fifty thousand square yards of dirt and rock smothered intertidal marine life. Large numbers of rock sole, Pacific cod, and rockfish were killed by the shock wave. After the evacuation, about three thousand otters remained on Amchitka before the hydrogen bomb. When Jim Estes visited the island a few months after the blast, he counted only one hundred and fifty-five. "The blast killed a lot of otters," he said. The carcasses littering the beach showed that the otters had been killed by pressure waves. Their skulls had been fractured by the force of the blast, driving their eyeballs through the bones behind their sockets. Many otters just disappeared. Only about one in ten survived.

IN A WORLD lit by nuclear invention, Cormac McCarthy wrote, history is "a rehearsal for its own extinction." The *Bulletin of the Atomic Scientists* currently has the Doomsday Clock at ninety seconds to midnight. Yet, of the many legacies of the Atomic Age—the Manhattan Project, the bombings of Hiroshima and Nagasaki, and the nuclear arms race—the return of sea otters to their historic range has to count as one of the brightest.

Amchitka also had a big impact on the field of ecology. Jim Estes was a graduate student at Washington State at the time. A statistical ecologist at the AEC hired him to monitor the sea otters before and after the Cannikin blast. On some days, he counted otters from the shore; on others, he did aerial surveys from a military helicopter, flying about a hundred and fifty feet above the sea. The hardest job for Estes was capturing the marine mammals and radio-collaring them—it was hard for the otters too, but they didn't have to worry about getting their fingers chewed off. Fifty years later, Estes still has the scars. He was counting and following the otters, but he was also preoccupied with finding a PhD project while he was on Amchitka. "The challenge was figuring out exactly what to do," he later wrote. "At the time, I knew almost nothing about the ocean and very little about ecology."

Working on the island before the detonation, Estes had considered a project examining the influence of habitat on sea otters. Kelp productivity was high around Amchitka, supporting sea urchins. The presence of the otters' favorite food would help explain the marine mammal's abundance. A classic bottom-up study—more kelp equals more urchins equals more otters. But when Bob Paine, a professor at the University of Washington, visited Amchitka, he suggested a radically different approach. Paine had published articles on the role of sea stars in preventing monopolies of mussels and other prey species; the leggy predators boosted diversity by opening up space for other organisms, such as seaweed and chitons. "Rather than wondering how the kelp forests affected otters (a view Bob found to be obvious and uninteresting), why not explore how otters affected the kelp forests?" Estes recalled forty-five years later.

So in the early 1970s, Estes followed Paine's advice, traveled the Aleutians, and tracked the differences between islands with plenty of otters, like Amchitka, and otterless islands. He noticed that the landscapes were physically similar—sheer cliffs, black sands, grasslands—but things were strikingly different beneath the ocean surface. The islands with otters had extensive kelp forests, mostly ungrazed areas of brown algae, with sparse populations of sea urchins, barnacles, and mussels. The otterless islands were almost kelp-free; in place of algae, there were vast mussel beds, chitons, barnacles, and lots of sea urchins.

What was going on? Estes proposed that the otters were structuring the nearshore communities through a process that came to be known as a trophic cascade; in the absence of otters, sea urchin abundance climbed. (We discussed this process for wolves and spiders earlier in the book. Estes's early work, published in *Science,* was foundational for these later studies.) Urchins eat kelp fronds that fall to the ocean floor, but in the absence of predators, they will emerge from the refuge of cracks and crevices and actively forage on living kelp, eating the holdfasts that tether the algae to the sea bottom. Once the kelp is cut loose, it dies. Thus, the echinoderms create urchin barrens, vast areas of invertebrates and hardly any kelp.

By eating the urchins, otters reduce consumption of the macroalgae, allowing the kelp forests to thrive. On the islands with otters, giant kelp towered hundreds of feet above the ocean floor, creating a nursery and a home for hundreds of species of fish and invertebrates, a rainforest of the sea. The presence of otters changed everything.

Estes's work on this iconic marine mammal would never have happened without Amchitka. His research had been funded by the Atomic Energy Commission before the blast, and the otter population recovered after the explosion. "The AEC did a lot of things that moved learning and conservation forward," he said. "They didn't do it out of the goodness of their hearts. They did it because it was becoming the politically and socially correct thing to do." The Cannikin explosion catalyzed his career, much as Surtsey had done for Erling and Borgthór half the world away. The translocations—intended to save an at-risk population from gruesome atomic deaths—would provide insight into how restoring a single species can result in ecological transformation. Someone else might have studied the otters and made these discoveries, Estes mused, or maybe not. "Who knows how the whole field [of trophic ecology] would have progressed?"

It certainly would have looked different for the marine ecologists of the Pacific Northwest. It seems like every other month a paper about sea otters drops in a leading journal, whether about their role as climate engineers (*Science*), the part they play in sea cow extinction (*Proceedings of the National Academy of Sciences*), or their aphrodisiac effects on the sex life of seagrasses (also *Science*).

One far-flung mammal almost hunted to extinction in a small part of the world became an iconic species for how animals shape the world.

Two sea otters the size of tractor tires stared me down as I waited to cross the street in Sitka.

Their adorable faces adorned the side of a passing bus on its way to a cruise ship, shilling for a local tour company. More than two hundred years after Sitka was the center of the fur trade, otters are still money.

There were several small boats that offered two-hour tours to see the otters, puffins, and whales of Sitka Sound. In this city that had once survived by killing otters, many people were now employed in shipping tourists out to watch otters or fish for salmon, halibut, and cod. I took one of those boats out onto Sitka Sound. Its ultimate destination, after we saw a few puffins and humpback whales, was Black Rock, a nursery for female sea otters and their pups. Like most charismatic animals, including many beloved pets, sea otters are natural charmers—they have those huge, brown-eyed stares, big mustaches of white whiskers, black triangle snouts, and ample curiosity, spending lots of time grooming at the ocean surface.

As we pulled up to the rock, one mother yelped.

"'Where is Junior?'" our tour guide translated.

Floating on her back, she had her head up and her rear paddle-like feet in the air, like a dark sail extended just above the surface. She was surrounded by kelp, as if the textbook lesson of a trophic cascade had been set for our arrival.

The mother screamed. And then the pup responded with a high-pitched, gull-like *I'm here, I'm here, I'm here.* To my ears, it all sounded a bit panicked, but these calls might be normal baby talk in the otter world.

They reunited, the pup swimming onto Mom's lap.

Things settled down on the rock. It was grooming time. An otter's life is eat, rest, groom. They groom for more than two hours a day, doing somersaults and rubbing and pleating their fur. They feed for about eight hours a day. The smallest marine mammals, typically weighing about seventy pounds or so, the size of a Labrador retriever, sea otters don't have blubber, so they rely on their thick fur and high metabolism—they have a core temperature of 101 degrees—to survive in the cold waters of the North Pacific. A dog might eat 2.5 percent of its body weight a day; otters are eating machines, consuming ten times that, more than a quarter of their body weight daily. (I'm sure our retriever would be up to the challenge if we would just open the pantry door.) Otters rely on ATP (there's that phosphate again) for energy, like the rest of us, but their mitochondria, the cellular workhorses, are leakier than most. They use this lost energy to heat their bodies. We can do this, too, though to a lesser degree—consider how working out warms you up in the winter.

There's been a lot of pooping and dying in this book, as there should be in any examination of animals' influence on ecosystems, but eating is just as important, carving out entire ecosystems, as sea otters do. If, as the Dalai Lama once noted, death is like changing our clothes when they are worn and old, eating is a way to maintain the wardrobe for as long as we can.

Or are we truly wearing the same wardrobe? As I watched the otters of Black Rock, all of them children of the bomb, descendants of the otters transported by Goose and Hercules in the sixties and seventies, I thought of my own origins: made in New York City, raised on Long Island. No matter where I go, it's hard not to feel like there is some physical marker, something bone-deep, connecting me to my hometown. How much of my New York–built bones and organs were on that boat, floating out there on Sitka Sound? Ninety percent? Fifty percent?

As it turns out, it was probably much closer to zero. Our skeletons are replaced about once a decade (technically, the change is 10 percent per year). Fat cells last about eight years. Our stomach lining turns over about once a week. We are continually taking in elements and minerals and releasing them back into the atmosphere, into the waterways, and into the soils of the lithosphere. Chimeras of our lives, we are constructed and mortared according to our movements and our diets. Walking along Allah Creek in Alaska, I had watched the lives of salmon slip away. As I passed on the carcasses to Daniel Schindler and George Pess to extract the otoliths, it was sometimes hard to see which side of life, or death, each fish was on, their bodies breaking down almost before they died.

We tend to think of death as an end, but your body replaces cells all the time. Yet the neurons in the central nervous system—the brain and spinal cord—remain the same across a lifetime.

"That's what I love about these body cells," said the neuron to the synapse. "I get older, they stay the same age."

In the decades since Deppa and his colleagues released 517 sea otters in southeast Alaska, the North Pacific's otter population has grown to about 125,000 individuals, and they have become the engineers of an entire coastline. Estes's early observations of the ecological impact of sea otters in the Aleutians have played out in real time and across much larger regions. Urchin barrens once stretched through vast areas, from Alaska to Washington State. After the Aleutian airlift, many of those barrens were transformed into thick canopies of kelp. Between 1988 and 2003, otters removed ninety-nine out of a hundred urchins from Sitka Sound. Kelp forests increased by more than 99 percent in the region. The forests provide food and shelter for more than eight hundred species, including sea lions, harbor seals, lingcod, gobies, moray eels, octopuses, crabs, sea anemones, and brittle stars. Lobsters, otters, sea stars, and all kinds of animals excrete nitrogen into the surrounding waters. The kelp can take up these nutrients and those released by humans. That doesn't happen in an urchin barren.

There have been plenty of surprises too, things that even Estes couldn't predict. As the otters' range expanded along the coast, fish became more abundant in those same regions, attracting bald eagles to the new kelp forests. After otters returned to Pleasant Island in Glacier Bay, wolves appeared. The deer population crashed on the island, and instead of leaving, the canids switched to scavenging otters, allowing them to persist on the island in the absence of deer—another marine subsidy.

Sadly, it's too late for sea otters and kelp to restore one ancient relationship, one that was severed soon after Bering and Steller sailed through the Aleutians. The sea cow that Steller described went extinct in 1768, just twenty-seven years after he saw it. Steller's sea cow was never hunted commercially, but sailors killed it for food while harvesting otters. It was the first documented extinction of a marine animal. In 2016, Estes and his colleagues showed that it wasn't just harpoons that killed the sea cows. With the otters hunted out, urchins increased and kelp declined. The buoyant herbivorous sea cow was left at the surface, starving as its food disappeared. Steller was perhaps the only European naturalist ever to see a living sea cow in the North Pacific—and he precipitated its demise.

WHEN THE ALEUTIAN otters were first delivered to southeast Alaska, they ate sea urchins almost exclusively. But as the urchin barrens declined, the otters turned their attention to gumboots (or chitons), crabs, abalone, and geoduck clams. Geoducks, at more than two pounds apiece, are among the most valuable commercial invertebrate fisheries in the region. Otters, like people, dive for clams to depths of up to sixty feet or so. In the 1990s, fishermen around Sitka started to get nervous. "These fucking sea otters are going to take over the sound," one resident said when he first saw an otter. "Before you know it, they will be in town, in the harbor, in front of the house. They are going to eat every clam and abalone for miles."

With close-cropped silver hair and gray eyes, Mike Miller, a member of the Sitka tribe, was born just before Deppa released the otters in Sitka. He rarely saw otters growing up, but in the 1990s when he was working on a tugboat, he recalled seeing something unusual when he was running his boat up north of Sitka. "In the flat calm days, we'd see these big things on the radar that weren't supposed to be there," he told me while we sat in Jan Straley's marine mammal lab at the University of Alaska, not far from Sitka Harbor. He looked out for rocks and other hazards, but the unidentified floating object turned out to be a big raft of male otters. "I was shocked at the numbers up there."

When otter numbers started to take off in the 1990s, "people said, 'The world's going to end,'" Miller recalled. "A lot of people in the shellfish industry made career choices to leave Sitka." But some people thought that otter hunting could be an answer. In the United States, it is illegal to hunt sea otters, or any other marine mammal, under the Marine Mammal Protection Act of 1972, but Alaska Natives retain the right to hunt otters, whales, and seals under a co-management agreement between the tribes and the federal government. Indigenous-led otter hunts were initially focused on restoring local invertebrates such as crabs and clams.

"In the late nineties, early two thousands, we looked at ways to promote economic development and make sure it didn't get abused as just an otter-slaughter mechanism," Miller said. Funding became available as the timber harvest in southeast Alaska started to decline and local fish plants closed. Classes to make sea otter handicrafts started up—Natives can't sell untreated otter pelts outside the tribe—and a tannery was opened for otter furs.

In the years that followed, Sitka tribal members started hunting more otters. Pelt prices were high, around two hundred fifty dollars each, and there was growing interest in traditional handicrafts. As otters declined, the local invertebrates became more common. "We think that the increased otter harvest, which was up to forty to fifty percent of the population per year, had a direct impact on the resurgence of the important subsistence invertebrates," like Dungeness crabs and geoducks. You don't have to eradicate otters to get an ecological response, but hunting pressure clearly made a difference. "We are rewriting the story," Miller said.

I had visited the Sitka otters before, during a research trip in 2014. There were otter rafts—what a lovely collective noun—of more than twenty bachelor males floating in the ale-brown algae. They kept their eyes on our boat, seemingly without a care in the world. They seemed present in a BBC Earth *Blue Planet* kind of way, comfortable in their own furs as we watched them through binoculars or took pictures through telephoto lenses.

The few males I saw on my return trip eight years later were porpoising away from our small moving boat, skipping on the surface, then taking long dives. They reminded me of the brown bears at Nerka running uphill or into the woods at our approach. It's not unusual when you foray into the natural world to see animals from an oblique angle—especially if they're accustomed to being hunted. According to Miller, otters avoid the small open skiffs that hunters use but are more relaxed around the larger tourist boats that stop to take pictures. "It's amazing how smart they are," he said.

As I see it, Miller's idea, which he calls the Sitka effect, folds humans into the ecosystem, rather than seeing us as an outside perturbation. If done in a responsible way—one that allows for abundant sea otters in some areas and managed hunts in others—it sounds a lot like the traditional Native approach. "A lot of people have no use for otters," Miller said. "It's too bad, because if they understood the benefits of kelp forests and the role of otters in creating fish nurseries, they might change their mind."

CLEARLY, THE RETURN of wildlife comes with tradeoffs. In nearby British Columbia, the Cannikin airlift prompted a new otter economy, one that continues to grow. Researchers at the University of British Columbia have valued the services that would be provided by the full recovery of otters on Vancouver Island at forty million dollars a year. The estimate includes tourism dollars—people would pay about thirty-one million a year to see otters. The marine mammals would enhance fisheries for lingcod and other species that use kelp forests as nurseries by about seven million dollars per year. They might even play an important role in the carbon cycle: Kelp forests use carbon dioxide to grow. When kelp dies, some of it is eaten, the carbon released back into the atmosphere. But some of it is also exported out to the deep sea, where carbon can be locked away for decades. This doesn't happen in an urchin barren. Carbon sequestration by otter-assisted kelp forests has been valued at more than $1.6 million a year. But there can be losses too. The decline in fisheries for Dungeness crabs, mussels, geoduck clams, and other bottom-dwelling invertebrates would cost fishers on Vancouver Island about $5.5 million a year in lost catch.

In Sitka, otter hunting has brought pelts to the community for sale while expanding the number of abalone, gumboots, and sea urchins. It has even attracted people to the Sitka tribe. "There are younger people who would like to hunt otters or own otter pelts," Miller said. There are discussions about increasing the tribe by lowering the quarter-blood level—having one Tlingit grandparent is the current standard—to something more inclusive. Miller is proud of the Sitka effect—the balance of harvesting otters, the return of benthic invertebrates, and a respect for the marine mammals as essential to Tlingit culture and heritage. But he was cautious, too, concerned that an expansion of the otter hunt could result in a population crash and a backlash against marine managers.

"It's like docking a boat," Miller said. "You can dock it safely ten thousand times, but the first time you break a piling, everybody remembers that."

In 1953, Eugene Odum wrote in his classic *Fundamentals of Ecology* that his fellow biologists should be able to show that it's as much fun to repair the biosphere as it is to fix a radio or the family car. Have those analogies lost their meaning entirely? These days we toss out broken radios (if we ever had them) and bring the computerized car to the shop. What about a planet in desperate need of repair? We need to get to net-zero carbon emissions as quickly and equitably as possible. But even that won't stop climate change. Many geoengineers say that tinkering with the Earth—especially the carbon cycle—is a terrible idea. But some say that the idea of not doing anything, of letting the oceans acidify and the land overheat, is even worse.

The restoration of wildlife populations might be one of the best nature-based tools we have to face the climate crisis. Wild animals, through their movements and behaviors—their eating, pooping, and dying—can help rebuild ecosystems, recycle and redistribute nutrients, keep the planet a little cooler, and address the biodiversity crisis. Up to a million animal and plant species are threatened with extinction, including one out of every ten birds, one out of four mammals, and one out of three sharks and rays. Throughout much of our species' time on Earth, the human impact on wild animals has been about reduction: in numbers, body size, range, and migratory patterns. That broad pattern is still mostly true, with a few exceptions.

Since the mid-twentieth century, there have been several national and international pushes to prevent the extinction of wildlife. Many of these policies, including the Endangered Species Act and treaties designed to reduce wildlife trafficking, have the laudable and much-needed goal of preventing extinction. At times, they've done more than stop the killing; they've promoted risky recovery efforts, like removing all the California condors from the wild, breeding them in captivity, fostering chicks with puppets, and releasing the offspring back into native areas where they can be left alone at last. But these laws often don't go far enough. If we are to revitalize the planet—to restore its animal heartbeat—we will need thriving populations of wild mammals, birds, amphibians, reptiles, fish, insects, crustaceans, and other invertebrates.

What's a reasonable if lofty goal? In *Half-Earth*, E. O. Wilson proposed designating 50 percent of Earth's surface as a nature reserve. Can we envision a world where half or even two-thirds of the planet's mammals and birds are wild? It's a long shot, especially if people are still eating meat on the bone and the population is increasing from eight billion to nine billion or ten. As fast as the number of humans is growing, the numbers of cattle, chickens, and sheep are increasing even faster.

How do we return to a land of giants? The most straightforward way to start is to protect the wilderness we have, places like Surtsey, the Hawaiian Islands Humpback Whale National Marine Sanctuary, Yellowstone National Park, the Maasai Mara, and the many

Restoring wild animals so that they are the majority of mammalian biomass will help fight the biodiversity crisis, support nature-based climate solutions, and revitalize the movement of nitrogen and phosphorus around the world.

parks and refuges that surround Bristol Bay in Alaska. When we were sitting beside Pick Creek, Schindler noted: "Some of the pieces of habitat that are important are tiny—the size of a square meter, in some cases, and ranging all the way up to the river valleys, which are hundreds of square kilometers. From a conservation standpoint, we usually focus on the big stuff, but the effects we have are mostly on the small stuff." The trouble is, we don't measure how important a small stream could be to an entire population of salmon or bears. Every scrap of land, every river, every hunk of coral, counts.

Expanding conservation efforts to areas that are currently unprotected can reverse biodiversity loss and stabilize the climate. Conservation scientist Eric Dinerstein and his colleagues have proposed a "global safety net" that targets conserving biodiversity, enhancing carbon storage, and restoring animal movement via wildlife and climate corridors. The safety net relies on expanding protected lands from the current state of about 15 percent to 46 percent, roughly in line with Wilson's ideas. Large intact wilderness areas and regions that protect the last populations of endangered species are given high priority. Restoration efforts can also benefit wildlife and carbon goals; community forestry programs in Nepal, aimed at large, highly degraded mid-elevation forests, have doubled the forest cover in twenty-four years, storing about three hundred million tons of carbon.

Look, there's no way around it. We have spoken about the benefits of eating—by bison, spiders, and sea otters, among others—but we're going to have to consume less, consume better, if we're going to rewild the world. A managed retreat is the only reasonable approach. Remember how the loss of mammoths helped start an ice age? More than 15 percent of the planet's greenhouse gases comes from raising meat, which requires far more energy and land than processing fruits and vegetables. By reducing our consumption of animals, and thus the vast grazing allotments and croplands that are dedicated to growing food for cattle, pigs, sheep, chickens, and other domestics, we could draw down carbon emissions and reduce the artificial fertilizers essential to their feed. Leah Garcés, of Mercy for Animals, and others call for a transition away from factory farms. Her Transfarmation Project has looked at several crops that could be grown in converted chicken warehouses, including cucumbers, strawberries, tomatoes, and mushrooms. Others see precision fermentation—a process that uses genetically modified microbes to produce meat and dairy products—as the future of animal proteins.

Some wildlife managers and livestock producers propose a more holistic type of grazing, where cattle have a smaller carbon hoofprint and wildlife share the pastures. This is sometimes referred to as a land-sharing approach. If we make ranches and farms more wildlife-friendly, wild animals will have more room to roam and provide ecological functions, like redistributing nutrients across landscapes. Others suggest that it's better to intensify food production while setting aside more land for wild animals and native plants—a land-sparing strategy.

"Do you concentrate or do you spread out?" Andrew Balmford, a conservation scientist at Cambridge University, asked during a Zoom call. His work shows that high-yield agriculture provides more food for people and opens more space for wildlife. Rather than trying to create wildlife-friendly farms with lower yields, as good as that feels, he believes it's more productive to use some farmlands intensively while taking other agricultural lands out of production and restoring grasslands and forests. Balmford's work demonstrates that land-sparing efforts, including habitat restoration, are cheaper for taxpayers. This approach increases the yield on farms and achieves biodiversity and carbon targets.

Balmford even challenges that sacred cow of many environmentalists: organic farming. Hedgerows and bee-friendly flowers sound great, but they aren't all that effective for wildlife conservation. Organic farms, he noted, typically achieve poorer yields, in part because they rely on organic sources of nutrients, like green manures from legume plants, and restrict the use of herbicides and antibiotics. They're also expensive. "There's a lot that needs to be changed with conventional high-yield agriculture," he told me. "But we can't feed the planet on organic food."

It's hard to love intensive farming, with its boxed-in animals, whiffs of ammonia, antibiotics, and animal-welfare issues, especially when compared to, say, the bucolic farms of Vermont, where landscape fragmentation is often seen as a thing of beauty: the pasture, the hay bales, the copse on the hill turning Life Saver colors in the fall, a mountain partially cleared. Many of my colleagues push back against land-sparing for reasons rooted in health and aesthetics and concern for animal welfare. "Conservationists who loathe the idea of sparing in an agricultural context for gut reasons, which I understand," Balmford said, chuckling, "have no trouble at all thinking that's what you ought to do with tourists," making certain parts of parks accessible and others hard to reach. These are all fair points, though in much of the United States and Europe, retired agricultural land is developed, not spared. But if we don't take advantage of this transition to more efficient agriculture now, it will be hard to get the land back.

WE HAVE GONE beyond the planetary boundaries for nitrogen, phosphorus, carbon, and biodiversity loss. We have too many of these nutrients in some places, where they run off from farms and lawns, prompting harmful algal blooms, and too little elsewhere, reducing productivity. Can animals help? To restore the world so that animals truly matter—as ecosystem engineers, sources of nutrient subsidies, and providers of daily wonder like the weather—we'll need *critterscapes*—areas created by abundant and free-roaming animals rather than by humans. This approach requires land-sparing and land-sharing.

One of the reasons that rewilding works is that it takes advantage of how animals eat, poop, and die and how they reproduce. It scales up because of the nature of biological processes such as population growth and expansion. I recently joined Chris Doughty, who came up with the idea of the animal circulatory system, and his postdoc Roo Abraham to calculate how rewilding could revitalize the phosphorus cycle. When whales and other marine mammals feed at depth and relieve themselves at the surface, they move nutrients vertically through their poop and pee. Seabirds move nutrients from offshore to islands and other coastal lands, as we saw at Surtsey. Salmon and other fishes bring marine nutrients upriver, through carcasses, poop, and pee. Predatory bears, scavengers, and insects move these nutrients around. As the seasons change, migrating bison and other large animals disperse nutrients as they feed in grasslands and choreograph the green wave. Finally, there's a world of insects, global capillaries, moving nutrients across the landscape.

How can we restore these animal subsidies? Taxes on mined phosphate or voluntary phosphorus credits, purchased like carbon credits, could fund biodiversity projects that would restore animal pathways. In the Amazon, for example, the restoration of large herbivores such as tapirs, peccaries, and howler monkeys could move about $900 million worth of phosphorus each year. Across the world, rewilding could increase the transport of phosphorus tenfold. Although our study looked only at vertebrates, the movement of this essential element by insects and other small mammals would certainly boost the numbers.

When we think of animals as part of the critterscape, we can get beyond the land-sharing and land-sparing debate. If modernity began when we left the farm, as English writer John Berger argued, then perhaps postmodernity, or postindustrial ecology, will begin when we value animals more for the services they provide—carbon mitigation, nutrient subsidies, recreation, awe—than for the products we extract from them, such as meat, milk, and fur.

We need sharing. We need sparing. We need connections between the two. Amanda Subalusky noted that fortress conservation, with protected areas isolated from human disturbance, won't necessarily work in a place like the Maasai Mara but that a new approach built on the early resistance of the Maasai people to colonial development could emerge. "The Maasai were integrated into this ecosystem for hundreds of years," she noted, and they resisted colonial development in the area. Indigenous groups around the world—from the Great Plains to the African savanna to the Amazon—have been caretakers of animals and their movements for hundreds of generations. The future of wildlife conservation could extend this relationship beyond the borders of Indigenous lands.

To restore the natural movement of animals, we'll need to address the disruption in migration by fences, roads, dams, and towns. After the slaughter of tens of millions of bison in North America, for instance, there were too few left to make any meaningful migration. Today, they couldn't travel very far even if they wanted to. Vast, impenetrable barriers restrict their movements.

"The way forward is to map out these animal corridors with detailed maps and GPS movement information," Matt Kauffman said after we discussed his work on bison and wolves in Yellowstone. "We could use that info to guide where we develop roads, railroads, and fences, where we put in holes: underpasses and overpasses." I think of this infrastructure as road-sharing through better design, but we'll also need more road-sparing by protecting and increasing roadless areas, making them less accessible to human development. Technological solutions, using GPS and drones, can help keep landscapes open if we know where the animals need to go. "That work is happening," Kauffman said, "but there needs to be a lot more of it."

ONE SUCH EFFORT aims to create a network of federal lands across the American West to restore two iconic species and diminish the hoofprint of another. The Western Rewilding Network focuses on restoring beavers and wolves across almost 200,000 square miles of contiguous land. The American beaver, perhaps the most renowned ecosystem engineer, helps protect watersheds and surrounding lands by slowing and storing water behind dams. Each fall, beavers gather riverside vegetation for the winter, collecting sticks and stones to reinforce their lodge walls. In the process, they alter river and stream systems. About two hundred million of these aquatic rodents once lived in the area that is now the Lower Forty-Eight. Wetlands created by beavers might have covered 10 percent of the total land area. After Europeans colonized the West, nine out of every ten beavers were hunted, trapped, or displaced.

Where they exist in the wild, beavers improve water quality, reduce flash-flood risks downstream of their dams, and create wildlife and fish habitats. Beaver dams can limit the flow of nitrogen and phosphorus out to lakes and estuaries, helping to curb harmful algal blooms and nutrient loss. The wood and sediment that collect in the soggy meadows behind dams can sequester carbon for hundreds of years. (But like sea otters, it should be noted, beavers don't always make perfect neighbors: their dams can also flood fields, roads, and tree plantations that people prefer to keep dry.) To boost their populations, we would probably have to reintroduce beavers to many watersheds via truck, boat, or parachute, since they don't move far. But once they take hold, little would need to be done. Beavers, like sea otters, are sticky—if humans don't overhunt them.

As we saw in Yellowstone, gray wolves have returned to the West after decades of absence, following restoration efforts and endangered species protections. But they're still found in only 14 percent of their historic range, with a total population of about 3,500, a fraction of the hundreds of thousands they numbered in the past. Montana continues to pass anti-wolf legislation, calling for the killing of 450 wolves with the clear intent of reducing the number of wild canids in the state. But there are hopeful signs: in 2020, voters in Colorado passed Proposition 114 to reintroduce wolves west of the Continental Divide, with plans to move about thirty to fifty of the wild canids from Montana and other western states.

Building beaver and wolf populations is relatively easy. Much more challenging is the reduction of another iconic animal of the West—cows. Cattle grazing can change the landscape, causing stream and wetland degradation, altering fire regimes, limiting the growth of woody species like willow, and degrading riparian buffers. On public lands in the West, grazing allotments cover more than 240,000 square miles, an area bigger than California—but they produce only about 2 percent of U.S. beef. Forty-four species protected under the Endangered Species Act are threatened by livestock grazing. The wildlife network is a land-sparing approach. If 30 percent of grazing allotments were retired, many species would benefit, taking a step toward ecological healing.

OPEN THE GATES. You never know what will happen when you take down the fences—whether built of barbed wire or human intolerance—lay down your guns, and let the animals run the place. Restoration and rewilding can be full of surprises.

In Zimbabwe, white rhinos, the lawn mowers of the savanna, were persecuted for decades in an effort to curb sleeping sickness. The parasitic disease—transmitted by the bite of an infected tsetse fly—can cause fatigue and headaches; untreated, it can lead to mental decline and death. Deprive the tsetse fly of a blood meal, it was thought, and you'll take out the disease. Rhinos, water buffalo, duikers, and other herbivores were gunned down, even after it was shown that insecticides were far more effective in controlling the flies and the disease. (The widespread use of pesticides, of course, might have negative knock-on effects on other invertebrates perceived as beneficial.) The loss of these plant-eaters changed the fire ecology of the savanna. The swards grew high, and the grasses went wild. A typical fire burned more than twelve hundred acres, several times larger than fires had in the past.

After the culls stopped in the 1970s, rhinos and other herbivores came charging back. They kept the grasses short. In the absence of fuel, the size of the typical savanna fire was reduced fiftyfold, to less than twenty-five acres. The more dung the researchers counted—a way to measure the number of rhinos without getting too close—the fewer the fires. Rhinos, elephants, and buffalo offered another anti-fire benefit: the paths they trampled through the vegetation created natural firebreaks. Meanwhile, their poop spread nutrients and seeds around the grasslands. The flesh and bones of dead herbivores—rhino and elephant falls—formed nutrient hot spots.

Rewilding projects can occur anywhere from small, fenced lawns to large landscapes like Yellowstone. In northern Denmark, a coalition of rewilding groups released seven European bison into a ten-thousand-acre enclosure of meadows and old agricultural fields. This keystone species is expected to boost the diversity of plants and animals by browsing, grazing, dispersing seeds, and transporting nutrients. In the Scottish Highlands, scientists released wild boar into several enclosures, ranging from an acre to several hundred, to examine the effects of rooting behavior on forests and bracken communities.

Tear down the fences, and animals could once again river the grasslands, forests, waters, and skies. Rewilding aspires to minimal human intervention. In places like the Swiss National Park, there is no hunting, no management, no agriculture. Brown bears and wolves were recently sighted for the first time in a century. There has also been some reseeding; ibex were introduced in the early twentieth century and bearded vultures in the 1990s. Hunting is allowed outside the park to help reduce conflicts between deer and neighboring landowners concerned about crops and trees.

The best-studied high-impact systems are often islands, which have their own natural barriers. For instance, the loss of giant tortoises in the Galápagos after humans arrived led to several plant extinctions, as wetlands created by the turtles disappeared. After giant tortoises were translocated back to the Galápagos and other small islands, they started dispersing large seeds and creating wallows that opened up new wetlands.

Many conservationists agree with animal restoration efforts, though when rewilding relies on ecological replacements—introducing a new species that would play a similar role as an extinct one—things get murkier. Can Eurasian horses help restore ancient ecosystem processes in North America that were once filled by native equids or are they just another invasive species burdening the native habitats? Time plays a role here. Support for rewilding among ecologists generally declines when a species or a close relative has been absent for more than five thousand years or so. Sea otters had been hunted out of southeast Alaska for less than a century before they were relocated. Their restoration wasn't controversial, at least among conservationists. The same is true for peregrine falcons, gray wolves, and many other recently extirpated species.

Still, the return of long-lost animals can capture the imagination. Elephants disappeared from Europe about thirty thousand years ago and are unlikely to be introduced outside of heavily fenced experimental areas. Morten Lindhard, a biologist with the Danish Nature Agency, wanted to give it a try, but the Copenhagen Zoo refused to lend him their elephants. So he turned to a traveling circus. In 2008, three elephants that had retired from performing were released for three days into a controlled area with a high density of birch and juniper west of Copenhagen. It wasn't much, as experiments go, but Lindhard saw evidence that the elephants had caused habitat disturbance—a good thing for biodiversity, as we've learned—and added, "The animals seemed to have a great time."

THE RIPPLE EFFECTS of the Atomic Age continue to this day. There are about fifty thousand descendants of the sea otters that were evacuated from Amchitka and nearby islands. That's around one out of every three otters in the entire North Pacific Ocean. The translocation saved hundreds of otters from the blast, launched one of the most successful cases of rewilding on the planet, restored vast kelp forests, and kick-started one of the defining concepts in ecology.

Amchitka is now a wildlife refuge. Critterscapes do not have to be pristine. Many remote islands on the Pacific Proving Grounds, where the United States tested more than twenty nuclear bombs, are now protected areas with thriving wildlife, though the craters from the detonations on places like the Bikini Atoll can still be seen from space. At least one other nuclear disaster appears to have benefited wildlife populations. After the Chernobyl accident in April 1986, the worst nuclear-power disaster in history, the Chernobyl Exclusion Zone was created on the border of Ukraine and Belarus. Though much of the abandoned land was highly contaminated with radioactive material, Chernobyl has become a refuge for wolves, lynx, and brown bears and a reintroduction site for European bison and the critically endangered Przewalski's horse. Like the Cannikin explosion, the Chernobyl disaster created an opportunity, and the area is an iconic natural experiment—two thousand square miles of fence-free land in the heart of central Europe. (Humans, it appears, are worse than radiation for many animals.) Can we catalyze widespread rewilding without the deep pockets of a military agency or a nuclear disaster? Well, there is that climate crisis, which is potentially far more explosive than any nuclear bomb.

Let's replace our carbon footprint with the tracks of wild animals.

Protecting and restoring wildlife can enhance carbon capture and storage in prairies, savannas, kelp forests, coral reefs, forests, and oceans through animals' diet, death, and feces. Os Schmitz, the grasshopper and zoogeochemistry guy, and colleagues have estimated that the protection and restoration of animals—animating the carbon cycle—could prompt an additional uptake of 6.4 billion tons of carbon dioxide each year. That's about a sixth of current annual global emissions.

At sea, the protection and return of fish, whales, sea otters, sharks, seabirds, and other animals are at the center of nature-based solutions to combat climate change: storing carbon in their bodies, sequestering it when they die in deep waters, protecting kelp forests, and boosting the growth of phytoplankton. On land, the restoration of large grazing animals could have three major impacts on the climate cycle: reducing the fuel for fires; increasing the albedo, or reflection of light, back into space by reducing the presence of dark leaves; and enhancing carbon storage in the soil. The return of grazing wildebeests to the Serengeti turned the savanna grasslands into a carbon sink by reducing the fuel for fires, a shift large enough to offset all of East Africa's annual fossil-fuel emissions. The restoration of bison on the Great Plains could increase uptake by 600 million tons of carbon dioxide each year. There will be methane. Many wild plant-eaters, such as bison, elk, and deer, emit the gas when they digest their food, but we can offset that with a modest reduction in livestock; cows and other animals raised for meat emit almost ten times more methane than wild herbivores do. Restoring the tracks and scats of wild animals is a natural climate solution to reduce warming.

We can make animals as vast and topical as the weather, with the predictability of spring and the drama and surprise of a thunderstorm. One blustery April morning, I walked out into the yard. Three squirrels were excavating the lawn. No doubt they'd miss a few acorns they had buried in the fall. Perhaps one oak will survive beyond the well-manicured edge of the grass. A chickadee perched in a sugar maple dropped a small white splat on my jacket collar. Two ravens flew overhead—joyous carcass-eaters—leaving a musical trail in the sky.

Will our legacy be drumsticks and broken wishbones? Or will whales, seabirds, salmon, bison, cicadas, and backyard birds be the lasting expression of who we are?

Acknowledgments

The idea for this book emerged during a Fulbright/National Science Foundation Arctic Research grant at the University of Iceland. The Gund Institute for Environment at the University of Vermont has been my intellectual home for years, providing support and inspiration as the project moved from proposal to manuscript. I finished the book during a fellowship at Harvard's Radcliffe Institute for Advanced Study, where I had few distractions aside from the dazzling creative and intellectual stimulation from my other fellows. I send my deep appreciation to the staff and fabulous colleagues of these institutions. Thank you for sharing your ideas, joy, and enthusiasm with me during the months of fieldwork, research, and writing.

For invaluable assistance in the field, thank you to Borgthór Magnússon, Bjarni Sigurdsson, and the rest of the wonderful crew at Surtsey; Brian and Ruth Bowen, Mark Hixon, and Chris Gabriele and Paul Berry in Hawaii; Daniel Schindler and the Alaska Salmon Program and Jan Straley in Sitka, Alaska; Dan Gruner in Maryland; Jeremy Kiszka and Alvaro Pereira in Florida; and Lauren McGarvey and Rick McIntyre in Yellowstone. When all hope of getting to Surtsey appeared to be lost, the Icelandic coast guard interrupted their fishing surveys to take us to the island. *Takk fyrir.*

For key conversations that appeared in the book, thanks to Andrew Balmford, Karen Bjorndal, Julia Cavicchi, Charlie Crisafulli,

Jerry Deppa, Chris Doughty, Jim Estes, Nick Graham, Claudio Gratton, Jim Helfield, Gordon Holtgrieve, David Hu, Matt Kauffman, Bun Lai, Kristin Laidre, Gary Lamberti, Leroy Little Bear, Mike Miller, Kim Nace, Bob Naiman, Erling Olafsson, George Pess, Jeff Pierce, Tom Quinn, Os Schmitz, Tatiana Schreiber, Victor Smetacek, Craig Smith, Amanda Subalusky, Jens Svenning, Rob Toonen, Rick Wallen, Pat Walsh, Kawika Winter, Louie Yang, and Patricia Yang.

For essential background and discussions, thanks to Roo Abraham, Lars Bejder, Rahul Bhatia, Jamie Botsch, Jenny Boylan, Corey Bronstein, Joseph Bump, Scott Collins, Chris Dutton, Brodie Fischer, Brendan Fisher, Amy Gulick, Jesse Hale, Philip Hamilton, Amy Knowlton, Jamie and Sue McCarthy, Kevin Miller, Dan Monson, Sarah Morley, Taylor Ricketts, Mark Rifkin, Marie Roman, Jenny Stern, Greg Streveler, Freydís Vigfúsdóttir, Nacho Villar, Skip Wallen, Jane Watson, and Taylor White. For help and hospitality in the field, thanks to Chris Boatright, Ray Hilborn, Kieko Matteson, Steph Matti, Diane Sweeney, and Frank Zelko.

Carolyn Savarese found a lovely home for this book in a tumultuous time. Thank you for looking after me from day one. Ian Straus helped shape this manuscript from its early stages to the final chapter. Tracy Behar enthusiastically brought the book over the finish line. Thanks to both and to Karen Landry and the fabulous team at Little, Brown. Thanks to Alex Boersma for the beautiful illustrations. Tracy Roe, the Serena Williams of copyediting, improved the book from the first page to the last.

Emma MacKenzie, my research partner at Radcliffe, was creative, enthusiastic, and frank, saving me from several blunders. Emma Wetsel at the University of Vermont helped me with this book and related whale-research projects. For reading drafts and chapters, thanks to Mark Hixon, David Hu, Borgthór Magnússon, Bill Patrick, Heidi Pearson, George Pess, Nate Sanders, and Amanda Subalusky. For early inspiration and stewardship, thanks to John Eisenberg, Elizabeth Kolbert, Jim McCarthy, Kathy Robbins, Ellen

Scordato, and Dave Wiley. Thanks to Debora Greger for her thousand and one great ideas and for sending me clippings, links, and books that made their way into these pages. I appreciate the friendship and research gifts from Paul and Lynn Lattanzio and Win and Alli Pescosolido.

This book was a collective effort. I met wonderful animals along the way, and back at home, my family provided emotional, intellectual, and logistical support. Laura Farrell read and reread several chapters. Nian Farrell-Roman helped with titles and artwork. Flo Roman provided love and curiosity. Zoey dragged me outside on the coldest mornings and hottest afternoons, reminding me that every day is worthy of tail-wagging joy and that our basic needs and local adventures form the rhythm of life. Thanks and love to all.

Key References

1. Beginnings

Croft, Betty, et al. "Contribution of Arctic Seabird-Colony Ammonia to Atmospheric Particles and Cloud-Albedo Radiative Effect." *Nature Communications* 7 (2016): 1–10.

Devred, Emmanuel, Andrea Hilborn, and Cornelia den Heyer. "Enhanced Chlorophyll-a Concentration in the Wake of Sable Island, Eastern Canada, Revealed by Two Decades of Satellite Observations: A Response to Grey Seal Population Dynamics?" *Biogeosciences* 18 (2021): 6115–32.

Fridriksson, Sturla. *Surtsey: Ecosystems Formed.* Reykjavík: University of Iceland, 2005.

Graham, Nicholas A. J., et al. "Seabirds Enhance Coral Reef Productivity and Functioning in the Absence of Invasive Rats." *Nature* 559 (2018): 250–53.

Magnússon, Borgthór, Sigurdur Magnússon, and Sturla Fridriksson. "Developments in Plant Colonization and Succession on Surtsey During 1999–2008." *Surtsey Research* 12 (2009): 57–76.

Magnússon, Borgthór, et al. "Plant Colonization, Succession and Ecosystem Development on Surtsey with Reference to Neighbouring Islands." *Biogeosciences* 11 (2014): 5521–37.

Magnússon, Borgthór, et al. "Seabirds and Seals as Drivers of Plant Succession on Surtsey." *Surtsey Research* 14 (2020): 115–30.

Thórarinsson, Sigurdur. *Surtsey: The New Island in the North Atlantic.* New York: Viking, 1964.

2. Deep Doo-Doo

Clements, Christopher F., et al. "Body Size Shifts and Early Warning Signals Precede the Historic Collapse of Whale Stocks." *Nature Ecology and Evolution* 1 (2017): 1–6.

Hutchinson, G. Evelyn. "The Biogeochemistry of Vertebrate Excretion." *Bulletin of the American Museum of Natural History* 96 (1950): 1–554.

Lane, Nick. *Transformer: The Deep Chemistry of Life and Death*. New York: W. W. Norton, 2022.

Lavery, Trish J., et al. "Iron Defecation by Sperm Whales Stimulates Carbon Export in the Southern Ocean." *Proceedings of the Royal Society B* 277 (2010): 3527–31.

Pearson, Heidi C., et al. "Whales in the Carbon Cycle: Can Recovery Remove Carbon Dioxide?" *Trends in Ecology and Evolution* 38 (2023): 238–49.

Pitman, Robert L., et al. "Skin in the Game: Epidermal Molt as a Driver of Long-Distance Migration in Whales." *Marine Mammal Science* 36 (2019): 565–94.

Quaggiotto, Martina, et al. "Past, Present and Future of the Ecosystem Services Provided by Cetacean Carcasses." *Ecosystem Services* 54 (2022): 101406.

Roman, Joe. *Listed: Dispatches from America's Endangered Species Act*. Cambridge, MA: Harvard University Press, 2011. (Brief sections of "Deep Doo-Doo" and "Heartland" were adapted from Joe Roman's *Listed*, *Whale*, and "Deep Doo-Doo: You Can Learn a Lot About a Whale from Its Feces," *New Scientist*, December 23, 2006.)

———. *Whale*. London: Reaktion, 2006.

Roman, Joe, et al. "Whales as Marine Ecosystem Engineers." *Frontiers in Ecology and the Environment* 12 (2014): 377–85.

Schmitz, Oswald J., et al. "Animals and the Zoogeochemistry of the Carbon Cycle." *Science* 362 (2018): eaar3213.

Smith, Craig R., et al. "Whale Fall Ecosystems: Recent Insights into Ecology, Paleoecology, and Evolution." *Annual Review of Marine Science* 7 (2015): 571–96.

3. Eat, Spawn, Die

Bouchard, Sarah S., and Karen A. Bjorndal. "Sea Turtles as Biological Transporters of Nutrients and Energy from Marine to Terrestrial Ecosystems." *Ecology* 81 (2000): 2305–13.

Helfield, James M., and Robert J. Naiman. "Effects of Salmon-Derived Nitrogen on Riparian Forest Growth and Implications for Stream Productivity." *Ecology* 82 (2001): 2403–9.

———. "Keystone Interactions: Salmon and Bear in Riparian Forests of Alaska." *Ecosystems* 9 (2006): 167–80.

Hilderbrand, Grant V., et al. "Role of Brown Bears (*Ursus arctos*) in the Flow of Marine Nitrogen into a Terrestrial Ecosystem." *Oecologia* 121 (1999): 546–50.

Holtgrieve, Gordon W., and Daniel E. Schindler. "Marine-Derived Nutrients, Bioturbation, and Ecosystem Metabolism: Reconsidering the Role of Salmon in Streams." *Ecology* 92 (2011): 373–85.

McLennan, Darryl, et al. "Simulating Nutrient Release from Parental Carcasses Increases the Growth, Biomass, and Genetic Diversity of Juvenile Atlantic Salmon." *Journal of Applied Ecology* 56 (2019): 1937–47.

Merz, Joseph E., and Peter B. Moyle. "Salmon, Wildlife, and Wine: Marine-Derived Nutrients in Human-Dominated Ecosystems of Central California." *Ecological Applications* 16 (2006): 999–1009.

Naiman, Robert J., et al. "Pacific Salmon, Marine-Derived Nutrients, and the Characteristics of Aquatic and Riparian Ecosystems." *American Fisheries Society Symposium* 69 (2009): 395–425.

Quinn, Thomas P., et al. "A Multidecade Experiment Shows That Fertilization by Salmon Carcasses Enhanced Tree Growth in the Riparian Zone." *Ecology* 99 (2018): 2433–41.

Quinn, Thomas P., et al. "Transportation of Pacific Salmon Carcasses from Streams to Riparian Forests by Bears." *Canadian Journal of Zoology* 87 (2009): 195–203.

Schindler, Daniel E., et al. "Pacific Salmon and the Ecology of Coastal Ecosystems." *Frontiers in Ecology and the Environment* 1 (2003): 31–37.

Tiegs, Scott D., et al. "Ecological Effects of Live Salmon Exceed Those of Carcasses During an Annual Spawning Migration." *Ecosystems* 14 (2011): 598–614.

Tonra, Christopher M., et al. "The Rapid Return of Marine-Derived Nutrients to a Freshwater Food Web Following Dam Removal." *Biological Conservation* 192 (2015): 130–34.

4. Heartland

Alerstam, Thomas, and Johan Bäckman. "Ecology of Animal Migration." *Current Biology* 28 (2018): R968–72.

Geremia, Chris, et al. "Migrating Bison Engineer the Green Wave." *Proceedings of the National Academy of Sciences* 116 (2019): 25707–13.

Heinrich, Bernd. *Life Everlasting: The Animal Way of Death*. New York: Mariner, 2013.

Lott, Dale F. *American Bison: A Natural History*. Berkeley: University of California Press, 2002.

Mueller, Natalie G., et al. "Bison, Anthropogenic Fire, and the Origins of Agriculture in Eastern North America." *Anthropocene Review* 8 (2021): 141–58.

Punke, Michael. *Last Stand: George Bird Grinnell, the Battle to Save the Buffalo, and the Birth of the New West*. New York: HarperCollins, 2009.

Ratajczak, Zak, et al. "Reintroducing Bison Results in Long-Running and Resilient Increases in Grassland Diversity." *Proceedings of the National Academy of Sciences* 119 (2022): e2210433119.

Subalusky, Amanda L., et al. "Annual Mass Drownings of the Serengeti Wildebeest Migration Influence Nutrient Cycling and Storage in the Mara River." *Proceedings of the National Academy of Sciences* 114 (2017): 7647–52.

Subalusky, Amanda L., et al. "The Hippopotamus Conveyor Belt: Vectors of Carbon and Nutrients from Terrestrial Grasslands to Aquatic Systems in Sub-Saharan Africa." *Freshwater Biology* 60 (2015): 512–25.

Subalusky, Amanda L., and David M. Post. "Context Dependency of Animal Resource Subsidies." *Biological Reviews* 94 (2019): 517–38.

Wenger, Seth J., Amanda L. Subalusky, and Mary C. Freeman. "The Missing Dead: The Lost Role of Animal Remains in Nutrient Cycling in North American Rivers." *Food Webs* 18 (2019): e00106.

5. Chicken Planet

Bar-On, Yinon, Rob Phillips, and Ron Milo. "The Biomass Distribution on Earth." *Proceedings of the National Academy of Sciences* 115 (2018): 6506–11.

Bennett, Carys E., et al. "The Broiler Chicken as a Signal of a Human Reconfigured Biosphere." *Royal Society Open Science* 5 (2018): 180325.

Cushman, Gregory T. *Guano and the Opening of the Pacific World: A Global Ecological History*. Cambridge: Cambridge University Press, 2013.

Doughty, Christopher E., Adam Wolf, and Yadvinder Malhi. "The Legacy of the Pleistocene Megafauna Extinctions on Nutrient Availability in Amazonia." *Nature Geoscience* 6 (2013): 761–64.

Erisman, Jan W., et al. "How a Century of Ammonia Synthesis Changed the World." *Nature Geoscience* 1 (2008): 636–39.

Flojgaard, Camilla, et al. "Exploring a Natural Baseline for Large-Herbivore Biomass in Ecological Restoration." *Journal of Applied Ecology* 59 (2022): 18–24.

Otero, Xosé L., et al. "Seabird Colonies as Important Global Drivers in the Nitrogen and Phosphorus Cycles." *Nature Communications* 9 (2018): 246.

Sandom, Christopher, et al. "Global Late Quaternary Megafauna Extinctions Linked to Humans, Not Climate Change." *Proceedings of the Royal Society B* 281 (2014): 20133254.

Smith, Felisa A., Scott M. Elliott, and Kathleen S. Lyons. "Methane Emissions from Extinct Megafauna." *Nature Geoscience* 3 (2010): 374–75.

Smith, Felisa A., et al. "Body Size Downgrading of Mammals over the Late Quaternary." *Science* 360 (2018): 310–13.

Storey, Alice A., et al. "Radiocarbon and DNA Evidence for a Pre-Columbian Introduction of Polynesian Chickens to Chile." *Proceedings of the National Academy of Sciences* 104 (2007): 10335–39.

Worster, Donald. *Nature's Economy: A History of Ecological Ideas*. 2nd ed. Cambridge: Cambridge University Press, 1994.

Wulf, Andrea. *The Invention of Nature: Alexander von Humboldt's New World*. New York: Knopf, 2015.

6. Everybody Poops—and Dies

Berendes, David M., et al. "Estimation of Global Recoverable Human and Animal Faecal Biomass." *Nature Sustainability* 1 (2018): 679–85.

De Frenne, Pieter, et al. "Nutrient Fertilization by Dogs in Peri-Urban Ecosystems." *Ecological Solutions and Evidence* 3 (2022): e12128.

Doughty, Caitlin. "If You Want to Give Something Back to Nature, Give Your Body." *New York Times*, December 5, 2022.

Rosen, Julia. "Humanity Is Flushing Away One of Life's Essential Elements." *Atlantic*, February 8, 2021.

Wald, Chelsea. "The Urine Revolution: How Recycling Pee Could Help to Save the World." *Nature* 602 (2022): 202–6.

Yang, Patricia J., et al. "Duration of Urination Does Not Change with Body Size." *Proceedings of the National Academy of Sciences* 111 (2014): 11932–37.

Yang, Patricia J., et al. "Hydrodynamics of Defecation." *Soft Matter* 13 (2017): 4960–70.

7. Beach Read

Bahr, Keisha D., Paul L. Jokiel, and Robert J. Toonen. "The Unnatural History of Kāneʻohe Bay: Coral Reef Resilience in the Face of Centuries of Anthropogenic Impacts." *PeerJ* 3 (2015): e950.

Grupstra, Carsten G. B., et al. "Fish Predation on Corals Promotes the Dispersal of Coral Symbionts." *Animal Microbiome* 3 (2021): 1–12.

Roman, Joe, et al. "Lifting Baselines to Address the Consequences of Conservation Success." *Trends in Ecology and Evolution* 30 (2015): 299–302.

8. The Singing Tree

Kaup, Maya, Sam Trull, and Erik F. Y. Hom. "On the Move: Sloths and Their Epibionts as Model Mobile Ecosystems." *Biological Reviews* 96 (2021): 2638–60.

Yang, Louie H. "Periodical Cicadas as Resource Pulses in North American Forests." *Science* 306 (2004): 1565–67.

9. Cloudy with a Chance of Midges

Dreyer, Jamin, et al. "Quantifying Aquatic Insect Deposition from Lake to Land." *Ecology* 96 (2015): 499–509.

Einarsson, Rasmus, et al. "Crop Production and Nitrogen Use in European Cropland and Grassland 1961–2019." *Scientific Data* 8 (2021): 288.

Godfrey-Smith, Peter. *Metazoa: Animal Life and the Birth of the Mind*. New York: Farrar, Straus and Giroux, 2020.

Gratton, Claudio, Jack Donaldson, and M. Jake Vander Zanden. "Ecosystem Linkages Between Lakes and the Surrounding Terrestrial Landscape in Northeast Iceland." *Ecosystems* 11 (2008): 764–74.

LaBarge, Laura R., et al. "Pumas *Puma concolor* as Ecological Brokers: A Review of Their Biotic Relationships." *Mammal Review* 52 (2022): 360–76.

Schmitz, Oswald J. "Effects of Predator Hunting Mode on Grassland Ecosystem Function." *Science* 319 (2008): 952–54.

10. The Otter and the H-Bomb

Abraham, Andrew J., Joe Roman, and Christopher E. Doughty. "The Sixth R: Revitalizing the Natural Phosphorus Pump." *Science of the Total Environment* 832 (2022): 155023.

Balmford, Andrew. "Concentrating vs. Spreading Our Footprint: How to Meet Humanity's Needs at Least Cost to Nature." *Journal of Zoology* 315 (2021): 79–109.

Bown, Stephen R. *Island of the Blue Foxes: Disaster and Triumph on the World's Greatest Scientific Expedition.* New York: Da Capo Press, 2017.

Collas, Lydia, et al. "The Costs of Delivering Environmental Outcomes with Land Sharing and Land Sparing." *People and Nature* 5 (2023): 228–40.

Dinerstein, Eric, et al. "A 'Global Safety Net' to Reverse Biodiversity Loss and Stabilize Earth's Climate." *Science Advances* 6 (2020): eabb2824.

Estes, James A. *Serendipity: An Ecologist's Quest to Understand Nature.* Berkeley: University of California Press, 2016.

Estes, James A., and John F. Palmisano. "Sea Otters: Their Role in Structuring Nearshore Communities." *Science* 185 (1974): 1058–60.

Estes, James A., et al. "Trophic Downgrading of Planet Earth." *Science* 333 (2011): 301–6.

Gorra, Torrey R., et al. "Southeast Alaskan Kelp Forests: Inferences of Process from Large-Scale Patterns of Variation in Space and Time." *Proceedings of the Royal Society B* 289 (2022): 20211697.

Gregr, Edward J., et al. "Cascading Social-Ecological Costs and Benefits Triggered by a Recovering Keystone Predator." *Science* 368 (2020): 1243–47.

Jones, Ryan Tucker. *Empire of Extinction: Russians and the North Pacific's Strange Beasts of the Sea, 1741–1867.* New York: Oxford University Press, 2014.

Kinney, Donald J. "The Otters of Amchitka: Alaskan Nuclear Testing and the Birth of the Environmental Movement." *Polar Journal* 2 (2012): 291–311.

Kristensen, Jeppe A., et al. "Can Large Herbivores Enhance Ecosystem Carbon Persistence?" *Trends in Ecology and Evolution* 37 (2022): 117–28.

Malhi, Yadvinder, et al. "The Role of Large Wild Animals in Climate Change Mitigation and Adaptation." *Current Biology* 32 (2022): R181–96.

Perino, Andrea, et al. "Rewilding Complex Ecosystems." *Science* 364 (2019): eaav5570.

Ripple, William J., et al. "Rewilding the American West." *BioScience* 72 (2022): 931–35.

Schmitz, Oswald J., et al. "Trophic Rewilding Can Expand Natural Climate Solutions." *Nature Climate Change* (2023). doi.org/10.1038/s41558-023-01631-6.

Shin, Yunne-Jai, et al. "Actions to Halt Biodiversity Loss Generally Benefit the Climate." *Global Change Biology* 28 (2022): 2846–74.

Svenning, Jens-Christian, et al. "Science for a Wilder Anthropocene: Synthesis and Future Directions for Trophic Rewilding Research." *Proceedings of the National Academy of Sciences* 113 (2016): 898–906.

Index

Note: Italic page numbers refer to illustrations.

About the Author

JOE ROMAN IS a conservation biologist, marine ecologist, and editor 'n' chef of eattheinvaders.org. Winner of the 2012 Rachel Carson Environment Book Award for *Listed: Dispatches from America's Endangered Species Act*, Roman has written for the *New York Times*, *Science*, *Audubon*, *New Scientist*, *Slate*, and other publications. Like many of the animals he studies, Roman is a free-range biologist. He has worked at Harvard University, Duke University Marine Lab, University of Iceland, University of Havana, the U.S. Environmental Protection Agency, and the University of Vermont, where he is a fellow and writer in residence at the Gund Institute for Environment. **www.joeroman.com**